GULF WAR and HEALTH

UPDATED LITERATURE REVIEW OF DEPLETED URANIUM

Committee on Gulf War and Health:
Updated Literature Review of Depleted Uranium

Board of Population Health and Public Health Practice

INSTITUTE OF MEDICINE
OF THE NATIONAL ACADEMIES

THE NATIONAL ACADEMIES PRESS
Washington, D.C.
www.nap.edu

THE NATIONAL ACADEMIES PRESS 500 Fifth Street, NW Washington, DC 20001

NOTICE: The project that is the subject of this report was approved by the Governing Board of the National Research Council, whose members are drawn from the councils of the National Academy of Sciences, the National Academy of Engineering, and the Institute of Medicine. The members of the committee responsible for the report were chosen for their special competences and with regard for appropriate balance.

This study was supported by Contract V101(049A3)-P-0066 between the National Academy of Sciences and the Department of Veterans Affairs. Any opinions, findings, conclusions, or recommendations expressed in this publication are those of the authors and do not necessarily reflect the view of the organizations or agencies that provided support for this project.

International Standard Book Number-13: 978-0-309-11919-1
International Standard Book Number-10: 0-309-11919-7

Additional copies of this report are available from the National Academies Press, 500 Fifth Street, NW, Lockbox 285, Washington, DC 20055; (800) 624-6242 or (202) 334-3313 (in the Washington metropolitan area); Internet, http://www.nap.edu.

For more information about the Institute of Medicine, visit the IOM home page at **www.iom.edu.**

Printed in the United States of America

The serpent has been a symbol of long life, healing, and knowledge among almost all cultures and religions since the beginning of recorded history. The serpent adopted as a logotype by the Institute of Medicine is a relief carving from ancient Greece, now held by the Staatliche Museen in Berlin.

Suggested citation: IOM (Institute of Medicine). 2008. *Gulf war and health: Updated literature review of depleted uranium.* Washington, DC: The National Academies Press.

"Knowing is not enough; we must apply.
Willing is not enough; we must do."
—Goethe

INSTITUTE OF MEDICINE
OF THE NATIONAL ACADEMIES

Advising the Nation. Improving Health.

THE NATIONAL ACADEMIES

Advisers to the Nation on Science, Engineering, and Medicine

The **National Academy of Sciences** is a private, nonprofit, self-perpetuating society of distinguished scholars engaged in scientific and engineering research, dedicated to the furtherance of science and technology and to their use for the general welfare. Upon the authority of the charter granted to it by the Congress in 1863, the Academy has a mandate that requires it to advise the federal government on scientific and technical matters. Dr. Ralph J. Cicerone is president of the National Academy of Sciences.

The **National Academy of Engineering** was established in 1964, under the charter of the National Academy of Sciences, as a parallel organization of outstanding engineers. It is autonomous in its administration and in the selection of its members, sharing with the National Academy of Sciences the responsibility for advising the federal government. The National Academy of Engineering also sponsors engineering programs aimed at meeting national needs, encourages education and research, and recognizes the superior achievements of engineers. Dr. Charles M. Vest is president of the National Academy of Engineering.

The **Institute of Medicine** was established in 1970 by the National Academy of Sciences to secure the services of eminent members of appropriate professions in the examination of policy matters pertaining to the health of the public. The Institute acts under the responsibility given to the National Academy of Sciences by its congressional charter to be an adviser to the federal government and, upon its own initiative, to identify issues of medical care, research, and education. Dr. Harvey V. Fineberg is president of the Institute of Medicine.

The **National Research Council** was organized by the National Academy of Sciences in 1916 to associate the broad community of science and technology with the Academy's purposes of furthering knowledge and advising the federal government. Functioning in accordance with general policies determined by the Academy, the Council has become the principal operating agency of both the National Academy of Sciences and the National Academy of Engineering in providing services to the government, the public, and the scientific and engineering communities. The Council is administered jointly by both Academies and the Institute of Medicine. Dr. Ralph J. Cicerone and Dr. Charles M. Vest are chair and vice chair, respectively, of the National Research Council.

www.national-academies.org

COMMITTEE ON GULF WAR AND HEALTH: UPDATED LITERATURE REVIEW OF DEPLETED URANIUM

DAVID G. HOEL, PhD *(Chair),* Distinguished University Professor, Medical University of South Carolina, Charleston, SC

MICHAEL ASCHNER, PhD, Gray E.B. Stahlman Professor of Neuroscience, Vanderbilt University, Nashville, TN

MELISSA D. BEGG, ScD, Professor of Clinical Biostatistics and Director of Academic Programs, Mailman School of Public Health of Columbia University, New York

H. TIM BORGES, PhD, Research Staff Member, Oak Ridge National Laboratory, Oak Ridge, TN *(resigned February 2008)*

VIVIEN W. CHEN, MPH, PhD, Professor and Director of Epidemiology Program, School of Public Health, Louisiana State University Health Sciences Center, New Orleans, LA

HAROLD I. FELDMAN, MD, MSCE, Professor of Medicine and Epidemiology, University of Pennsylvania School of Medicine, Philadelphia, PA

PHILIP HARBER, MD, MPH, Professor and Chief, Division of Occupational and Environmental Medicine, David Geffen School of Medicine, University of California, Los Angeles

PATRICK J. HEAGERTY, PhD, Professor of Biostatistics, University of Washington School of Public Health and Community Medicine, Seattle, WA

KIYOUNG LEE, MPH, ScD, Assistant Professor, University of Kentucky College of Public Health and Seoul National University School of Public Health, Republic of Korea

JONATHAN LINKS, PhD, Professor and Director, Center for Public Health Preparedness, Johns Hopkins Bloomberg School of Public Health, Baltimore, MD

STAFF

ABIGAIL E. MITCHELL, PhD, Senior Program Officer
PETER JAMES, Senior Program Associate (through May 2007)
A. WEZI MUNTHALI, MPH, Senior Program Associate (through November 2007)
DEEPALI M. PATEL, Senior Program Associate (through November 2007)
JENNIFER E. SAUNDERS, MPP, MPH, Senior Program Associate (from November 2007)
RENEE WLODARCZYK, Research Assistant
JOSEPH GOODMAN, Senior Program Assistant
DANIELLE K. STOLL, Intern
NORMAN GROSSBLATT, Senior Editor
ROSE MARIE MARTINEZ, ScD, Director, Board on Population Health and Public Health Practice

Reviewers

This report has been reviewed in draft form by persons chosen for their diverse perspectives and technical expertise in accordance with procedures approved by the National Research Council's Report Review Committee. The purpose of this independent review is to provide candid and critical comments that will assist the institution in making its published report as sound as possible and to ensure that the report meets institutional standards of objectivity, evidence, and responsiveness to the study charge. The review comments and draft manuscript remain confidential to protect the integrity of the deliberative process. We wish to thank the following for their review of this report:

Odile David, University of Illinois College of Medicine
Fletcher Hahn, Lovelace Respiratory Research Institute, Scientist Emeritus
Robert Herrick, Department of Environmental Health, Harvard School of Public Health
Howard Hu, University of Michigan Schools of Public Health and Medicine
Ronald L. Kathren, Washington State University, Professor Emeritus
William N. Rom, Division of Pulmonary and Critical Care Medicine, New York University School of Medicine
Michael J. Thun, Epidemiology and Surveillance Research, American Cancer Society
Bailus Walker, Jr., Howard University Medical Center
Walter Willett, Department of Nutrition, Harvard School of Public Health
Judith T. Zelikoff, Department of Environmental Medicine, New York University School of Medicine

Although the reviewers listed above have provided many constructive comments and suggestions, they were not asked to endorse the conclusions or recommendations, nor did they see the final draft of the report before its release. The review of this report was overseen by **David J. Tollerud,** School of Public Health and Information Sciences, University of Louisville, and **Johanna T. Dwyer,** Tufts University School of Medicine and Friedman School of Nutrition Science, Tufts-New England Medical Center. Appointed by the Institute of Medicine and the National Research Council, respectively, they were responsible for making certain that an independent examination of this report was carried out in accordance with institutional procedures and that all review comments were carefully considered. Responsibility for the final content of this report rests entirely with the authoring committee and the institution.

Contents

TABLES AND BOXES

(Note that the tables for Chapters 3, 7, and 8 are at the ends of the chapters)

Tables

Boxes

Summary

The 1990-1991 Persian Gulf War was considered a military success, with few injuries or deaths. However, a number of veterans began experiencing symptoms—such as fatigue, cognitive difficulties, and sleep disturbances—after their return home. In response to growing concern about possible exposure to a biologic, chemical, or physical agent as the cause of the symptoms, Congress passed two laws in 1998: PL 105-277, the Persian Gulf War Veterans Act, and PL 105-368, the Veterans Programs Enhancement Act. Under the legislation, the Department of Veterans Affairs (VA) was directed to ask the Institute of Medicine (IOM) to evaluate the scientific literature regarding associations between illness and exposure to specific toxic agents, environmental or wartime hazards, or preventive medicines or vaccines related to Gulf War service.

In 1998, IOM began a program to examine health risks posed by specific agents and hazards to which Gulf War veterans might have been exposed during their deployment. Five reports have examined health outcomes related to depleted uranium, pyridostigmine bromide, sarin, and vaccines; insecticides and solvents; fuels, combustion products, and propellants; infectious diseases; and physiologic, psychologic, and psychosocial effects of deployment-related stress. A sixth IOM report examined the current health status of Gulf War–deployed veterans compared with their nondeployed counterparts. The present report updates the review of depleted uranium presented in the 2000 IOM report, *Gulf War and Health, Volume 1: Depleted Uranium, Pyridostigmine Bromide, Sarin, Vaccines* (hereafter referred to as *Volume 1*).

The Gulf War marked the first time that depleted-uranium munitions and armor were extensively used by the US military. Depleted uranium is used by the US military for both offensive and defensive purposes. Heavy-armor tanks have a layer of depleted-uranium armor to increase protection. Offensively, depleted uranium is used in kinetic-energy cartridges and ammunition rounds. The Army used an estimated 9,500 depleted-uranium tank rounds during the Gulf War. Ammunition containing depleted uranium was used in Bosnia-Herzegovina in 1994-1995 and in Kosovo in 1999; about 10,800 depleted-uranium rounds were fired in Bosnia-Herzegovina, and about 30,000 in Kosovo. Depleted-uranium–containing weapons also have been used in Operation Iraqi Freedom (OIF), which began in 2003. Because depleted uranium continues to be used by the military, the charge to IOM has been expanded to include not only veterans of the Gulf War but veterans returning home from OIF.

Military personnel have been exposed to depleted uranium as a result of friendly-fire incidents, cleanup and salvage operations, and proximity to burning depleted-uranium–containing tanks and ammunition. During the Gulf War, an estimated 134-164 people experienced "level I" exposure (the highest of three exposure categories as classified by the US Department of Defense) through wounds caused by depleted-uranium fragments, inhalation of airborne depleted-uranium particles, ingestion of depleted-uranium residues, or wound contamination by depleted-uranium residues. Hundreds or thousands more may have been exposed to lower exposure through inhalation of dust containing depleted-uranium particles and residue or ingestion from hand-to-mouth contact or contamination of clothing. Ten US military personnel who served in OIF had confirmed depleted uranium detected in their urine; all 10 had depleted-uranium embedded fragments or fragment injuries.

SUMMARY OF FINDINGS IN *VOLUME 1*

When *Volume 1* was published in 2000, few studies of health outcomes of exposure to depleted uranium had been conducted. Therefore, the committee studied the health outcomes of exposure to natural and processed uranium in workers at plants that processed uranium ore for use in weapons and nuclear reactors. After evaluating the literature, the committee concluded that there was inadequate or insufficient evidence to determine whether an association exists between uranium exposure and 14 health outcomes—lymphatic cancer, bone cancer, nervous system disease, reproductive or developmental dysfunction, nonmalignant respiratory disease, gastrointestinal disease, immune-mediated disease, effects on hematologic measures, genotoxic effects, cardiovascular effects, hepatic disease, dermal effects, ocular effects, and musculoskeletal effects. It also concluded that there was limited or suggestive evidence of *no* association between uranium and clinically significant renal dysfunction and between uranium and lung cancer at cumulative internal doses lower than 200 mSv.

CHARGE TO THE COMMITTEE

Since *Volume 1* was published in 2000, a number of studies of health outcomes of exposure to natural and depleted uranium have been published. For that reason and because depleted uranium continues to be used by the military, VA asked IOM to update the 2000 report and to take into consideration information published since *Volume 1*. In response, IOM entered into a contract with VA to conduct the following study:

> An IOM committee will review, evaluate, and summarize the scientific literature regarding the association between exposure to depleted uranium and long-term human health outcomes. The study committee will incorporate literature published since the 2000 IOM report *Gulf War and Health, Volume 1: Depleted Uranium, Pyridostigmine Bromide, Sarin, Vaccines* was written. The committee will make determinations on the strength of the evidence of associations between exposure to depleted uranium and human health outcomes.

THE COMMITTEE'S APPROACH TO ITS CHARGE

The committee began its evaluation by presuming neither the existence nor the absence of adverse health outcomes associated with exposure to depleted uranium. It sought to characterize and weigh the strengths and limitations of the available evidence. The committee did not concern itself with policy issues, such as decisions regarding disability, potential costs of compensation, or any broad policy implications of its findings.

An extensive search of the scientific literature generated about 3,500 titles and abstracts. After examination of the titles and abstracts to identify articles that appeared to be relevant to the committee's task (that is, articles on health outcomes of exposure to uranium), about 1,000 articles—including epidemiologic, toxicologic, and exposure-assessment studies—remained in the committee's reference database. Additional information was obtained from invited experts and the public during a meeting held on June 28, 2007, in Washington, DC.

After securing the full text of the articles mentioned above, the committee had to determine which ones would be included in the review.

For an epidemiologic study to be included in the committee's review, it had to be published in a peer-reviewed journal or to have undergone an equally rigorous process. A study also needed to be judged as methodologically sound, on the basis of inclusion of details of its methods, use of appropriate control or reference groups, statistical adjustment to control for confounders and minimize selection bias, and appropriate assessment of uranium exposure in the study population. It needed to examine long-term health outcomes and had to have a followup time sufficient to detect a relevant clinical effect. Finally, it had to include a relevant study population, that is, uranium-exposed workers, military personnel deployed to the Gulf War, or people who lived near a uranium-processing facility (uranium exposure in such residents may be similar to low-level exposures of military personnel). Studies in uranium miners were not included in the committee's evaluation because several issues related to confounding substantially limited the usefulness of those studies.

The committee used the evidence in the scientific literature to draw conclusions about associations between exposure to depleted uranium and specific adverse health outcomes. Those conclusions are presented as categories of strength of association. The categories have been used in many previous IOM studies, and they have gained wide acceptance by Congress, government agencies, researchers, and veteran groups. The categories are summarized below.

- **Sufficient evidence of a causal relationship.** Evidence is sufficient to conclude that a causal relationship exists between the exposure to uranium and a specific health outcome in humans. The evidence fulfills the criteria for sufficient evidence of an association (below) and satisfies several of the criteria used to assess causality: strength of association, dose-response relationship, consistency of association, temporal relationship, specificity of association, and biological plausibility.
- **Sufficient evidence of an association.** Evidence is sufficient to conclude that there is an association. That is, a consistent association unlikely to be due to sampling variability has been observed between exposure to uranium and a specific health outcome in human studies that were free of severe bias and that controlled for confounding.
- **Limited/suggestive evidence of an association.** Evidence is suggestive of an association between exposure to uranium and a specific health outcome, but the body of evidence is limited by insufficient avoidance of bias, insufficient control for confounding, or large sampling variability.
- **Inadequate/insufficient evidence to determine whether an association exists.** Evidence is of insufficient quantity, quality, or consistency to permit a conclusion regarding the existence of an association between exposure to uranium and a specific health outcome in humans.
- **Limited/suggestive evidence of no association.** Evidence is consistent in not showing an association between exposure to uranium of any magnitude and a specific health outcome. A conclusion of no association is inevitably limited to the conditions, magnitudes of exposure, and length of observation in the available studies.

CONCLUSIONS

The committee drew on information from the many studies published since 2000 and from *Volume 1* and reached its conclusions by interpreting the new evidence in the context of the entire body of literature. Most of the evidence on health effects of exposure to uranium came from studies of workers in uranium-processing mills and other facilities, and the committee relied heavily on those studies in developing its conclusions. Also taken into consideration in the evaluation were studies of Gulf War veterans exposed to depleted uranium and of residential exposure to uranium. All those studies were valuable in drawing conclusions, but they also had limitations. For example, the number of exposed people in many of the studies was relatively small, and this decreased the statistical power to detect a small excess risk of disease. The period of followup in several studies might have been too short to detect some diseases that are typically characterized by long latency; this limitation is of particular concern with respect to studies of cancer outcomes. And assessment of exposure to uranium was inadequate in many of the studies reviewed by the committee.

On the basis of the available literature, the committee concluded that there is *inadequate/insufficient evidence to determine whether an association exists* between exposure to uranium and all the health outcomes examined: lung cancer, leukemia, lymphoma (Hodgkin lymphoma and non-Hodgkin lymphoma), bone cancer, renal cancer, bladder cancer, brain and other nervous system cancers, stomach cancer, prostatic cancer, testicular cancer, nonmalignant renal disease, nonmalignant respiratory disease, neurologic effects, reproductive and developmental effects, and several other health outcomes (cardiovascular effects, genotoxicity, hematologic effects, immunologic effects, and skeletal effects). The committee's conclusions on lung cancer and nonmalignant renal disease differ from those in *Volume 1* (see "Summary of Findings in *Volume 1*" above). With respect to lung cancer, the committee decided not to place quantitative limits on the dose, primarily because of the wide variety of exposure-assessment methods used in the studies reviewed and the uncertainty in measurement of uranium exposure. With respect to nonmalignant renal disease, the committee decided that it could not rule out the occurrence of a renal effect "after exposure of any magnitude," as required to meet the definition of *limited/suggestive evidence of no association*.

In summary, the committee assigned the category *inadequate/insufficient evidence to determine whether an association exists* to each health outcome described above for one or more of the following reasons:

- Well-conducted studies showed equivocal results.
- The magnitude or frequency of a health outcome may be so low that it cannot be reliably detected given the sizes of the study populations.
- The available studies had limitations (such as inadequate exposure assessment or followup that was too short) that made it impossible to reach clear conclusions about health outcomes.

1

Introduction

On August 2, 1990, Iraqi forces invaded Kuwait. Five months later, the United States and its coalition allies launched an air offensive; and in February 1991, ground troops were deployed in a 4-day ground war. By April 1991, an official cease-fire was signed, and the last troops returned to the United States by June. In all, almost 700,000 troops had been deployed in the Persian Gulf War.

The war was considered a military success with few injuries or deaths, but a number of veterans began experiencing symptoms after their return, such as fatigue, cognitive difficulties, and sleep disturbances. In response to growing concern of possible exposure to a biologic, chemical, or physical agent as the cause of the symptoms, Congress passed two laws in 1998: PL 105-277, the Persian Gulf War Veterans Act, and PL 105-368, the Veterans Programs Enhancement Act. Those laws directed the Department of Veterans Affairs (VA) to task the Institute of Medicine (IOM) to evaluate the scientific literature regarding associations between illness and exposure to toxic agents, environmental or wartime hazards, or preventive medicines or vaccines associated with Gulf War service. They also provided a specific list of agents for IOM to review (see Box 1-1). Seven volumes have been published thus far: *Gulf War and Health, Volume 1: Depleted Uranium, Pyridostigmine Bromide, Sarin, Vaccines* (IOM, 2000); *Gulf War and Health, Volume 2: Insecticides and Solvents* (IOM, 2003); *Gulf War and Health: Updated Literature Review of Sarin* (IOM, 2004); *Gulf War and Health, Volume 3: Fuels, Combustion Products, and Propellants* (IOM, 2005); *Gulf War and Health, Volume 4: Health Effects of Serving in the Gulf War* (IOM, 2006); *Gulf War and Health, Volume 5: Infectious Diseases* (IOM, 2007a); and *Gulf War*

BOX 1-1
Agents Specified in PL 105-277 and PL 105-368

- The following organophosphorus pesticides:
  - Chlorpyrifos
  - Diazinon
  - Dichlorvos
  - Malathion
- The following carbamate pesticides:
  - Proxpur
  - Carbaryl
  - Methomyl
- Pyridostigmine bromide (used for nerve-agent prophylaxis)
- The following chlorinated hydrocarbons and other pesticides and repellents:
  - Lindane
  - Pyrethrins
  - Permethrins
  - Rodenticides (bait)
  - DEET (repellent)
- The following low-level nerve agents and precursor compounds at exposures below those which produce immediately apparent incapacitating symptoms:
  - Sarin
  - Tabun
- The following synthetic chemical compounds:
  - Mustard agents at exposures below those which cause immediate blistering
  - Volatile organic compounds
  - Hydrazine
  - Red fuming nitric acid
  - Solvents
- The following sources of radiation:
  - Depleted uranium
  - Microwave radiation
  - Radiofrequency radiation
- The following environmental particles and pollutants:
  - Hydrogen sulfide
  - Oil-fire byproducts
  - Diesel-heater fumes
  - Sand microparticles
- Diseases endemic to the region, including the following:
  - Leishmaniasis
  - Sand fly fever
  - Infections due to pathogenic *Escherichia coli*
  - Shigellosis
- Time-compressed administration of multiple live "attenuated" and toxoid vaccines

and Health, Volume 6: Physiologic, Psychologic, and Psychosocial Effects of Deployment-Related Stress (IOM, 2007b). The present report updates the review of depleted uranium that appeared in the first volume.

Since the passage of the two laws, the United States and its allies have deployed troops to the Persian Gulf region again, to Iraq (Operation Iraqi Freedom). Operation Iraqi Freedom began on March 29, 2003, with the arrival of US and British troops in Iraq, in an effort to remove Saddam Hussein from power. Although major combat operations ended on May 1, 2003, US and coalition troops continue to be deployed to Iraq to fight insurgency and assist with reconstruction and security. About 1.5 million US troops (active-duty military personnel and reservists) have been deployed to Operation Iraqi Freedom and Operation Enduring Freedom in Afghanistan (Report of the President's Commission on Care for America's Returning Wounded Warriors, 2007).

As discussed in the following chapter, depleted uranium is used by the US military for both offensive and defensive purposes. Exposure of US troops to depleted uranium can occur as a result of friendly-fire incidents, cleanup operations, and accidents. Exposure to depleted uranium was a reported concern among Operation Iraqi Freedom veterans participating in a retrospective review of veterans' health and exposure concerns (Helmer et al., 2007).

SUMMARY OF FINDINGS IN THE SECTION ON DEPLETED URANIUM IN *GULF WAR AND HEALTH, VOLUME 1*

When *Volume 1* was published in 2000, few studies on health outcomes of exposure to depleted uranium had been conducted. Therefore, the committee studied the health outcomes of exposure to natural and processed uranium in workers at plants that processed uranium ore for use in weapons and nuclear reactors. After evaluating the literature, the committee concluded that there was inadequate or insufficient evidence to determine whether an association exists between uranium and 14 health outcomes: lymphatic cancer, bone cancer, nervous system disease, reproductive or developmental dysfunction, nonmalignant respiratory disease, gastrointestinal disease, immune-mediated disease, effects on hematologic measures, genotoxic effects, cardiovascular effects, hepatic disease, dermal effects, ocular effects, and musculoskeletal effects. It also concluded that there was limited or suggestive evidence of no association between uranium and lung cancer at cumulative internal doses lower than 200 mSv and between uranium and clinically significant renal dysfunction.

THE DEPARTMENT OF VETERANS AFFAIRS REQUEST FOR THIS STUDY

Studies on health outcomes of exposure to uranium and depleted uranium published through 1999 were included in *Volume 1*. Since *Volume 1* was published in 2000, a number of new studies have been published. In addition, weapons systems that contain depleted uranium are being used in the military operations that began in Iraq in 2003, so there is potential for troop exposure to depleted uranium due to incidents of friendly fire and accidental fire. For those reasons, VA has asked IOM to update the 2000 report and to take into consideration information on health outcomes of exposure to depleted uranium that has been published since *Volume 1*.

THE COMMITTEE'S TASK

In response to VA's request, IOM entered into a contract with VA to conduct the following study: An IOM committee will review, evaluate, and summarize the scientific literature regarding the association between exposure to depleted uranium and long-term human health outcomes. The study committee will incorporate literature published since IOM's 2000 report, *Gulf War and Health, Volume 1: Depleted Uranium, Pyridostigmine Bromide, Sarin, and Vaccines*, was written. The committee will make determinations on the strength of the evidence of associations between exposure to depleted uranium and human health outcomes. The report might identify data gaps and subjects of scientific uncertainty and make recommendations for addressing them.

ORGANIZATION OF THIS REPORT

Chapter 2 gives information about the chemistry of uranium and its mode of action and about how depleted uranium is used by the US military. It also discusses radiologic and chemical mechanisms of uranium's action. The toxicology of uranium, including toxicokinetics and toxicodynamics, is summarized in Chapter 3. The committee's approach to its task is described in Chapter 4. Chapter 5 reviews available information on exposure to depleted uranium in military personnel and how it is detected in humans. The committee's rationale for selecting specific clinical end points is presented, with background information on each, in Chapter 6. Chapter 7 contains a comprehensive description of epidemiologic studies on health outcomes in populations exposed to depleted uranium or natural uranium. The final chapter, Chapter 8, presents the committee's conclusions.

REFERENCES

Helmer, D. A., M. Rossignol, M. Blatt, R. Agarwal, R. Teichman, and G. Lange. 2007. Health and exposure concerns of veterans deployed to Iraq and Afghanistan. *Journal of Occupational and Environmental Medicine* 49(5):475-480.

IOM (Institute of Medicine). 2000. *Gulf War and health, volume 1: Depleted uranium, sarin, pyridostigmine bromide, vaccines*. Washington, DC: National Academy Press.

———. 2003. *Gulf War and health, volume 2: Insecticides and solvents*. Washington, DC: The National Academies Press.

———. 2004. *Gulf War and health: Updated literature review of sarin*. Washington, DC: The National Academies Press.

———. 2005. *Gulf War and health, volume 3: Fuels, combustion products, and propellants*. Washington, DC: The National Academies Press.

———. 2006. *Gulf War and health, volume 4: Health effects of serving in the Gulf War*. Washington, DC: The National Academies Press.

———. 2007a. *Gulf War and health, volume 5: Infectious diseases*. Washington, DC: The National Academies Press.

———. 2007b. *Gulf War and health, volume 6: Physiologic, psychologic, and psychosocial effects of deployment-related stress*. Washington, DC: The National Academies Press.

Report of the President's Commission on Care for America's Returning Wounded Warriors. 2007. *Report of the president's commission on care for America's returning wounded warriors report*. http://www.pccww.gov/ (accessed October 13, 2007).

2

Background

Uranium is a radioactive element that occurs naturally in soil, rocks, surface and underground water, air, plants, and animals (ATSDR, 1999b). It is found at an average concentration of 0.0003% in Earth's crust and at 3.0 μg/L in seawater (Bleise et al., 2003). It also occurs in trace amounts in many foods and in drinking water as a result of its presence in the environment.

Uranium is the heaviest naturally occurring element. Its density is 19 times that of water and 1.65 times that of lead (Kirk, 1981; ATSDR, 1999b). The chemical symbol for uranium is U, and its Chemical Abstract Services Registry Number is 7440-61-1. It exists in nature as three isotopes, or forms. Isotopes have the same number of protons in the nucleus, and therefore are the same element, but a different number of neutrons. All three naturally occurring uranium isotopes are radioactive. The most abundant is ^{238}U (99.2745% abundance), and the second-most abundant is ^{235}U (0.7200%) (Lide, 1999). The natural abundance of ^{234}U is only 0.0055%. Uranium is not found in its elemental state but combined with other elements in about 150 known minerals (McDiarmid and Squibb, 2001).

The primary civilian use of uranium is as fuel for nuclear power plants (Cantaluppi and Degetto, 2000; Betti, 2003). Minute amounts are also used in the production of ceramic glazes, light bulbs, and photographic chemicals (ATSDR, 1999b). A person's daily intake of uranium is estimated to be 1-2 μg in food and 1.5 μg in each liter of water consumed (ATSDR, 1999b). The International Commission on Radiological Protection has reported that the average uranium content of the human body is 90 μg, including 69 μg in the skeleton and 7 μg in the kidneys (ICRP, 1975). A range of total body uranium of 2-62 μg has been noted in human postmortem studies (Wrenn et al., 1985).

Depleted uranium is a byproduct of the uranium enrichment process used to generate fuel for nuclear power plants. Depleted uranium is so named because it has been partially depleted of radioisotopes, the abundance of both ^{235}U and ^{234}U is lower than natural (it may also contain ^{236}U). The ratio of ^{238}U to ^{235}U in natural uranium is 137.88; in depleted uranium, it is 314.95 (Roth et al., 2003). The chemical properties of depleted uranium are the same as those of the enriched and natural forms (ATSDR, 1999b).

USES OF DEPLETED URANIUM

The need for enriched nuclear fuel has been present for decades, so depleted uranium, a byproduct of the enrichment process, is abundant and inexpensive. The chemical and physical properties of depleted uranium make it ideal for several military and commercial uses. It is 67% denser than lead (with a density of 18.9 g/cm^3), has a high melting point (2070°F, 1132°C), is highly pyrophoric, has a tensile strength comparable with that of most steels, and is chemically highly reactive (Kirk, 1981). It is used in commercial products, such as radiation shielding in medical equipment, aircraft counterweights, rotors, flywheels, ship ballasts, and gyroscopes (Cantaluppi and Degetto, 2000; Betti, 2003; Sztajnkrycer and Otten, 2004).

The US Army began researching the use of depleted uranium for military applications in the early 1970s (Bleise et al., 2003), and depleted uranium is now used both offensively and defensively. In the Gulf War, heavy-armor tanks had a layer of depleted-uranium armor to increase protection, and depleted uranium was used in kinetic-energy cartridges and ammunition rounds by the Army (105- and 120-mm tank ammunition), Air Force (armor-piercing munitions for the Gatling gun mounted on the A-10 aircraft), Marine Corps (Harrier aircraft and tank munitions), and Navy (rounds for the Phalanx Close-in Weapon System) (DOD, 2000). The Army used an estimated 9,500 depleted-uranium tank rounds during the Gulf War, many in training and practice (DOD, 2000).

The US military has continued to use depleted-uranium weapons. Ammunition containing depleted uranium was used in Bosnia-Herzegovina in 1994-1995 and in Kosovo in 1999 (Cantaluppi and Degetto, 2000; Bleise et al., 2003). According to North Atlantic Treaty Organization records, about 10,800 depleted-uranium rounds were fired in Bosnia-Herzegovina and about 30,000 in Kosovo (Bleise et al., 2003). Depleted-uranium–containing weapons have been used in Operation Iraqi Freedom (OIF), which began in 2003 (Burkart et al., 2005; NRC, 2008).

EXPOSURE OF MILITARY PERSONNEL TO DEPLETED URANIUM

The Gulf War marked the first time that depleted-uranium munitions and armor were extensively used by the US military (DOD, 2000). The Iraqi forces did not have such munitions. US military personnel were exposed to depleted uranium as a result of friendly-fire incidents, cleanup and salvage operations, and proximity to burning depleted-uranium–containing tanks and ammunition (DOD, 2000). Depleted-uranium–containing projectiles struck 21 occupied Army combat vehicles (15 Bradley fighting vehicles and 6 Abrams tanks) (AEPI, 1995). In addition, US forces used depleted-uranium rounds to destroy three unoccupied Abrams tanks to prevent them from being captured by the enemy, and five Abrams tanks became contaminated when depleted-uranium rounds were involved in onboard fires (AEPI, 1995). After the war, assessment teams and cleanup and recovery personnel may have had contact with depleted-uranium–contaminated vehicles or depleted-uranium munitions. In July 1991, a large fire occurred in Camp Doha near Kuwait City. This site housed a number of combat-ready vehicles, and the series of blasts and fires damaged or destroyed vehicles and munitions, including Abrams tanks and depleted-uranium munitions. Troops at the scene and those involved in cleanup efforts may have been exposed to depleted-uranium residue. Other troops may have been exposed through contact with vehicles or inhalation of depleted-uranium–containing dust.

In estimating the number of US personnel exposed to depleted uranium during the Gulf War and the extent of their exposure, the Department of Defense Office of the Special Assistant for Gulf War Illnesses categorized potential depleted-uranium exposure scenarios in three levels (DOD, 2000). The levels are described briefly below and in more depth in Chapter 5.

Level I, the highest exposure level, occurred in or near combat vehicles when they were struck by depleted-uranium rounds or when soldiers entered vehicles soon after impact. An estimated 134-164 people may have experienced level I exposure through wounds caused by depleted-uranium fragments, inhalation of airborne depleted-uranium particles, or ingestion of or wound contamination by depleted-uranium residues. Some 74 Gulf War veterans, including those with internal depleted-uranium fragments, are participating in the Depleted Uranium Follow-up Program, a medical surveillance followup study that began in 1993 at the Baltimore Veterans Affairs Medical Center (McDiarmid et al., 2007).

Level II, the intermediate exposure level, occurred when soldiers and civilian employees worked on depleted-uranium–contaminated vehicles or were involved in cleanup efforts from the Camp Doha fire. More than 700 people may have had level II exposure through inhalation of dust containing depleted-uranium particles and residue or through ingestion by hand-to-mouth contact or contamination of clothing.

Level III, the lowest level of exposure, occurred when troops were downwind of burning depleted-uranium ammunition or vehicles or of the Camp Doha fire or when personnel entered depleted-uranium–contaminated Iraqi tanks. These level III exposures could have occurred though inhalation or ingestion. Hundreds of people are thought to have experienced potential level III exposure, but there is little to substantiate the estimates.

The US Army conducted a study to model depleted-uranium aerosol exposures; the results of this study are presented in the "Capstone report" (USACHPPM, 2004). The exposure modeling characterized depleted-uranium aerosols as aerosols that would be generated by perforation of an Abrams tank or a Bradley fighting vehicle. Models were developed for level I, II, and III exposures. In addition, an evaluation of health outcomes of exposure to the depleted-uranium aerosols was conducted for level I inhalation exposures. Depending on the exposure scenario, the median intakes of depleted uranium range from 10 mg for a 1-min exposure in a ventilated Abrams tank with depleted-uranium armor to 710 mg for a 5-min exposure in an unventilated Abrams tank with depleted-uranium armor. The Capstone report is reviewed in detail in the National Research Council report *Review of Toxicologic and Radiologic Risks to Military Personnel from Exposure to Depleted Uranium During and After Combat* (NRC, 2008).

The Royal Society, which is the United Kingdom's equivalent of a national academy of science, convened an independent expert working group to review the evidence on health effects of exposure to depleted uranium. The Royal Society's "central estimate" (representative of the average person in the group of people exposed in that situation) for a level I inhalation exposure was 250 mg (Royal Society, 2001). The central estimates for level II and III exposures were 1-10 and 0.05-0.8 mg, respectively.

In a report prepared for the US Department of Energy, Marshall (2005) also estimated average exposures. The estimate for "nominal" level I inhalation exposure (representative of the average person in the group under study) was 250 mg, for level II exposure 40 mg, and for level III exposure 6 mg.

Military personnel potentially were exposed to depleted uranium during the Bosnia-Herzegovina and Kosovo wars (WHO, 2001). Aircraft-fired depleted-uranium munitions were used by the United States during those wars. Exposure would occur from handling munitions, from being protected by depleted-uranium–armored tanks, or after depleted-uranium use on the battlefield (Bolton and Foster, 2002). Urinary analyses have not found increased concentrations of uranium in several populations working in areas that might have been contaminated with depleted uranium: US National Guard troops deployed to Bosnia (May et al., 2004), German peacekeeping personnel serving in Kosovo (Oeh et al., 2007), and International Red Cross and Red Crescent Movement workers in Kosovo (Meddings and Haldimann, 2002).

As of September 30, 2007, 2,447 US military personnel who served in OIF had undergone a depleted-uranium bioassay (DOD, 2007). Ten of those personnel had confirmed urinary depleted uranium, and all ten had embedded fragments of depleted uranium or fragment injuries. Depleted-uranium concentrations were not found to be increased in 341 UK military personnel who were deployed to Iraq in 2003 (Bland et al., 2007).

RADIOLOGIC AND CHEMICAL EFFECTS OF EXPOSURE TO DEPLETED URANIUM

In considering the potential toxicity of depleted uranium, it is important to distinguish between radiologic and chemical toxic mechanisms. It is also important to note that the radiologic and chemical properties of uranium could act synergistically to cause health outcomes.

Radiologic Considerations

As discussed above, depleted uranium and naturally occurring uranium have different abundances of the three isotopes. The most notable difference is a decrease in the abundance of ^{235}U from 0.72% to 0.20% in depleted uranium, which reduces overall radioactivity by about 40% (Harley et al., 1999).

The radioactivity of a source is based on the number of radioactive atoms undergoing radioactive decay in a given period. Radioactive decay is the attempt of any atom to rearrange or transform the constituent protons and neutrons of its nucleus in such a way that the atom ends up having lower inherent energy. Radioactive decay occurs spontaneously because energy is given off, rather than consumed, in the process. The result of radioactive decay is an atom (the daughter) with less inherent energy than that which preceded it (the radioactive parent atom). Uranium isotopes decay to other radioactive elements that eventually decay to stable isotopes of lead (ATSDR, 1999b).

The term *radioactivity* describes how many radioactive atoms are undergoing radioactive decay every second. It does not reflect what type of radiation is being emitted or the energy of that radiation. The traditional unit of radioactivity is the curie (Ci); 1 Ci equals 3.7×10^{10} disintegrations per second (dps). A disintegration occurs when an atom undergoes radioactive decay. The International System unit of radioactivity is the Becquerel (Bq); 1 Bq is equivalent to 1 dps. Common units of measurement are summarized in Box 2-1.

Radioactive half-life is the amount of time it takes for radioactivity to decrease by half, that is, for half the radioactive atoms to undergo radioactive decay. The half-life of ^{238}U is 4.47×10^9 years, and the half-life of ^{235}U 7.04×10^8 years (Bleise et al., 2003); radioactivity never reaches zero but only keeps fractionally reducing.

The isotopes of uranium emit alpha particles. Alpha particles are positively charged ions composed of two protons and two neutrons. Because of their size and charge, alpha particles lose their kinetic energy quickly and have little penetrating power. The range of an alpha particle is about 4 cm in air and considerably less (25-80 μm) in tissue (ATSDR, 1999a). As a result, uranium is a radiation hazard mainly when uranium atoms are in the body. As noted above, uranium isotopes decay to other radioactive elements that eventually decay to stable isotopes of lead. In the decay process, beta particles and gamma rays are

BOX 2-1
Units of Measurement

Specific Activity
The **curie** (Ci) is the traditional unit of radioactivity defined as the quantity of any radioactive nuclide in which the number of disintegrations per second is 3.700×10^{10}. It is a concentration defined as the ratio of the amount of radioactivity divided by the mass or volume of radioactive substance. The International System unit of specific activity is the **becquerel** (Bq).

Absorbed Dose
The **gray** (Gy), formerly the rad, is the unit that describes the magnitude of absorbed radiation in terms of energy deposited in tissue. However, the amount of energy deposited in tissue does not account for differences in the biologic effects of different radiation types.

Dose Equivalent
The **rem** (roentgen-equivalent-man) is the traditional unit of measure that incorporates the relative biologic damage caused by different radiation types and deposition mechanisms. The International System unit for the biologically effective dose, dose equivalent, is the **sievert** (Sv).

	Specific Activity	Absorbed Dose	Biologically Effective Dose
Units	curie (Ci) becquerel (Bq)	gray (Gy) rad (old standard unit)	rem sievert (Sv)
Conversion	1 Bq = 1 transformation or disintegration per second = 2.7×10^{-11} Ci	1 Gy = 100 rad	1 mSv = 0.001 Sv 1 Sv = 100 rem

SOURCE: Adapted from ATSDR, 1999a,b.

emitted. Beta particles are high-energy electrons; the path length of a beta particle is up to 15 m in air and up to 1 cm in solids (ATSDR, 1999b). Gamma rays are electromagnetic ionizing radiation and constitute a radiation hazard even when present outside the body because they are highly penetrating.

Isotopes of uranium all have the same chemical properties because they all have the same number of protons, 92. However, variation in the number of neutrons gives the isotopes different radiologic properties. The radioactivity of isotopes can be compared by using specific activity, a measure of the number of nuclear transformations (disintegrations) per second per unit mass (see Box 2-1). The most abundant naturally occurring uranium isotope, ^{238}U, has the lowest specific activity (1.24×10^4 Bq/g) (AEPI, 1995). The high specific activity of

^{234}U (2.31×10^8 Bq/g) contributes about half the radioactivity of natural uranium, even though by weight its percentage is extremely small.

Estimates of radiation risk depend on the dose received by a person. Radiation dose is the amount of energy deposited per unit mass. The traditional unit of absorbed dose is the rad, which is defined as the absorption of 100 ergs/g. The International System unit of absorbed dose is the gray (Gy); 1 Gy equals 100 rad (see Box 2-1). To account for cellular and subcellular differences in energy deposition pattern by alpha particles, beta particles, and gamma rays, which affect biologic consequences, doses are often expressed as dose equivalents. The dose equivalent is the absorbed dose multiplied by a radiation weighting factor, which describes the ability of a given kind of radiation to produce a particular biologic effect relative to X-rays. Radiation weighting factors range from 1 (for X-rays) to over 20 (for some alpha particles). The traditional unit of dose equivalent is the rem. The International System unit is the sievert (Sv); 1 Sv equals 100 rem. In relating dose of ionizing radiation to risk, an extension of the dose equivalent is used to express dose as what would have been received if the whole body had been uniformly irradiated. The "effective dose equivalent" is the sum of dose equivalents to different organs or body tissues weighted in such a fashion as to provide a value proportional to radiation-induced somatic and genetic risk even when the body is not uniformly irradiated.

There are a number of radiologic-protection regulations and guidelines. The US Nuclear Regulatory Commission's regulations for occupational dose to individual adults state an annual limit of the total effective dose equivalent of 5 rem/year (50 mSv/year) (10 CFR 20.1201). The commission's regulations require that the total effective dose equivalent to individual members of the general public not exceed 0.1 rem (1 mSv) in a year exclusive of background radiation (10 CFR 20.1301). The background dose in the United States is about 0.36 rem/year.

Chemical Toxicity

As noted above, the enriched, natural, and depleted forms of uranium have identical chemical properties and therefore the same chemical toxicity. The chemical toxicity of a uranium compound depends on the nature of the compound, its solubility, and its route of exposure (inhalation, ingestion, or skin absorption). Chemical toxicity, characterized predominantly by renal dysfunction as a consequence of exposure to soluble uranium, and lung injury potentially caused by the ionizing radiation from uranium-decay isotopes are the best-characterized consequences of exposure to uranium compounds (Eidson, 1994). Relatively water-soluble compounds (uranyl nitrate hexahydrate, uranium hexafluoride, uranyl fluoride, uranium tetrachloride, and uranium pentachloride) are the most potent renal toxicants (ATSDR, 1999b). Sodium diuranate and ammonium diuranate, which are less water-soluble, are of moderate to low renal toxicity; and uranium tetrafluoride, uranium trioxide, uranium dioxide, uranium peroxide, and triuranium octaoxide, which are insoluble, have little potential to cause renal toxicity but could cause pulmonary toxicity if exposure is by inhalation (ATSDR, 1999b). Insoluble uranium compounds can remain in the pulmonary tissues, especially the pulmonary lymph nodes, for a long time and constitute a localized radiologic hazard. As a general rule, uranium in the intestinal tract is less readily absorbed than uranium from the respiratory tract and results in lower doses per unit intake.

DOSE-RESPONSE MODELING AND RISK ASSESSMENT

The committee was charged with evaluating the scientific literature on the effects of depleted uranium. As detailed in this report, the evaluation focused on direct experimental and observational evidence in animals and human populations. The committee acknowledges that there is a broader literature on risk assessment of radiologic and chemical toxicants, including uranium.

In general, population-based quantitative risk assessment is used in public health to inform intervention strategies, for example, in setting policy and regulations. Such risk assessment is not intended to estimate risk to any given individual in a population or to determine causality. Rather, it is intended to characterize population-attributable risk broadly to support population-level, not individual-level, decisions.

The current approach to quantitative risk assessment, developed by the National Research Council, consists of four steps: hazard identification, exposure assessment, dose-response modeling, and risk characterization (NRC, 1983, 1994). Of the four steps, the committee emphasized two as most relevant to its charge: hazard identification (that is, Does evidence of toxicity of depleted uranium exist at any level of exposure?) and exposure assessment (that is, What actual levels of exposure were experienced by military personnel serving in the Gulf War?). The committee considered mechanisms of both radiologic and chemical toxicity and a variety of cancer and noncancer outcomes or end points.

Elements of the risk assessment approach—notably dose-response modeling—vary among the cancer and noncancer end points. Cancer and genetic changes are modeled as a mathematical function in which risk increases with increasing exposure or dose. Although considerable controversy remains about the shape of the dose-response curve, especially at low doses, a linear no-threshold model has traditionally been used. This approach has been used for ionizing radiation as a carcinogen; for example, the National Research Council report *Health Risks from Exposure to Low Levels of Ionizing Radiation: BEIR VII Phase 2* endorses the use of such a model for radiogenic-cancer risk estimation (NRC, 2006).

A linear no-threshold dose-response model implies that cancer risk increases proportionally with increasing dose and that no "safe" dose exists (that is, every exposure or dose conveys some risk—low doses have low risk and higher doses proportionally higher risk). Such a model continues to be used for population-based quantitative risk assessment in public health, in spite of substantial uncertainties in its validity, because it is the most conservative or protective approach (that is, it yields the highest estimated risk for a given exposure or dose).

In the context of the committee's work, a key element of the examination of the possibility of adverse health effects of depleted uranium was the presence (or absence) of direct scientific evidence relevant to Gulf War veterans that could support the adoption of a no-threshold model for depleted-uranium cancer risk. The validity of the linear no-threshold model, especially for radiogenic cancer, is of greatest uncertainty at doses below 25 rem, the very range of doses to Gulf War veterans considered here. Thus, although a no-threshold model is used to estimate risk to a population, especially at higher doses, and would imply risk related to any level of depleted-uranium exposure, the committee chose to focus on direct evidence rather than a conservative, theory-driven approach in making its final determinations even while it remained mindful of the issues described here.

REFERENCES

AEPI (Army Environmental Policy Institute). 1995. *Health and environmental consequences of depleted uranium use in the U.S. Army*. Arlington, VA: U.S. Army Environmental Policy Institute.

ATSDR (Agency for Toxic Substances and Disease Registry). 1999a. *Toxicological profile for ionizing radiation*. Atlanta, GA: Public Health Service, Centers for Disease Control and Prevention.

———. 1999b. *Toxicological profile for uranium*. Atlanta, GA: Public Health Service, Centers for Disease Control and Prevention.

Betti, M. 2003. Civil use of depleted uranium. *Journal of Environmental Radioactivity* 64(2-3):113-119.

Bland, D. J., R. J. Rona, D. Coggon, J. Anderson, N. Greenberg, L. Hull, and S. Wessely. 2007. Urinary isotopic analysis in the UK armed forces: No evidence of depleted uranium absorption in combat and other personnel in Iraq. *Occupational and Environmental Medicine* 64:834-838.

Bleise, A., P. R. Danesi, and W. Burkart. 2003. Properties, use and health effects of depleted uranium (DU): A general overview. *Journal of Environmental Radioactivity* 64(2-3):93-112.

Bolton, J. P. G., and C. R. M. Foster. 2002. Battlefield use of depleted uranium and the health of veterans. *Journal of the Royal Army Medical Corps* 148(3):221-229.

Burkart, W., P. R. Danesi, and J. H. Hendry. 2005. Properties, use and health effects of depleted uranium. *International Congress Series* 1276:133-136.

Cantaluppi, C., and S. Degetto. 2000. Civilian and military uses of depleted uranium: Environmental and health problems. *Annali di Chimica* 90(11-12):665-676.

DOD (Department of Defense). 2000. *Environmental exposure report: Depleted uranium in the Gulf (II)*. Washington, DC: Department of Defense.

———. 2007. *Operation Iraqi Freedom depleted uranium bioassay results—sixth semiannual report and request for data submission*. Washington, DC: Department of Defense.

Eidson, A. F. 1994. The effect of solubility on inhaled uranium compound clearance: A review. *Health Physics* 67(1):1-14.

Harley, N. H., E. C. Foulkes, L. H. Hilborne, A. Hudson, and C. R. Anthony. 1999. *A review of the scientific literature as it pertains to Gulf War illnesses: Depleted uranium*. Santa Monica, CA: RAND.

ICRP (International Commission on Radiological Protection). 1975. *Report of the task group on reference man*. Oxford, UK: Pergamon Press.

Kirk, W. 1981. Depleted uranium. In *Mineral facts and problems*. Edited by W. Kirk. Washington, DC: Department of the Interior, Bureau of Mines. Pp. 997-1003.

Lide, D. R. 1999. *CRC handbook of chemistry and physics*. 80th ed. Edited by D. R. Lide. Boca Raton, FL: CRC Press.

Marshall, A. C. 2005. Analysis of uranium dispersal and health effects using a Gulf War case study. *Government Reports Announcements and Index* 23:212.

May, L. M., J. Heller, V. Kalinsky, J. Ejnik, S. Cordero, K. J. Oberbroekling, T. T. Long, K. C. Meakim, D. Cruess, and A. P. Lee. 2004. Military deployment human exposure assessment: Urine total and isotopic uranium sampling results. *Journal of Toxicology and Environmental Health A* 67(8-10):697-714.

McDiarmid, M. A., and K. S. Squibb. 2001. Uranium and thorium. In *Patty's toxicology*. 5th ed. Edited by E. Bingham, B. Cohrssen, and C. H. Powell. New York: John Wiley and Sons, Inc. Pp. 381-421.

McDiarmid, M. A., S. M. Engelhardt, M. Oliver, P. Gucer, P. D. Wilson, R. Kane, A. Cernich, B. Kaup, L. Anderson, D. Hoover, L. Brown, R. Albertini, R. Gudi, D. Jacobson-Kram, and K. S. Squibb. 2007. Health surveillance of Gulf War I veterans exposed to depleted uranium: Updating the cohort. *Health Physics* 93(1):60-73.

Meddings, D. R., and M. Haldimann. 2002. Depleted uranium in Kosovo: An assessment of potential exposure for aid workers. *Health Physics* 82(4):467-472.

NRC (National Research Council). 1983. *Risk assessment in the federal government: Managing the process*. Washington, DC: National Academy Press.

———. 1994. *Science and judgment in risk assessment*. Washington, DC: National Academy Press.

———. 2006. *Health risks from exposure to low levels of ionizing radiation: BEIR VII phase 2*. Washington, DC: The National Academies Press.

———. 2008. *Review of toxicologic and radiologic risks to military personnel from exposure to depleted uranium during and after combat*. Washington, DC: The National Academies Press.

Oeh, U., N. D. Priest, P. Roth, K. V. Ragnarsdottir, W. B. Li, V. Hollriegl, M. F. Thirlwall, B. Michalke, A. Giussani, P. Schramel, and H. G. Paretzke. 2007. Measurements of daily urinary uranium excretion in German peacekeeping personnel and residents of the Kosovo region to assess potential intakes of depleted uranium (DU). *Science of the Total Environment* 381(1-3):77-87.

Roth, P., V. Hollriegl, E. Werner, and P. Schramel. 2003. Assessment of exposure to depleted uranium. *Radiation Protection Dosimetry* 105(1-4):157-161.

Royal Society. 2001. *The health hazards of depleted uranium munitions: Part I*. London, UK: The Royal Society Working Group on the Health Hazards of Depleted Uranium Munitions.

Sztajnkrycer, M. D., and E. J. Otten. 2004. Chemical and radiological toxicity of depleted uranium. *Military Medicine* 169(3):212-216.

USACHPPM (US Army Center for Health Promotion and Preventive Medicine). 2004. *Capstone report: Depleted uranium aerosol doses and risks: Summary of U.S. Assessments*. Fort Belvoir, VA: US Army Heavy Metals Office, Chemical and Biological Defense Information Analysis Center.

WHO (World Health Organization). 2001. *Report of the world health organization depleted uranium mission to Kosovo*. Geneva, Switzerland: World Health Organization.

Wrenn, M. E., P. W. Durbin, B. Howard, J. Lipsztein, J. Rundo, E. T. Still, and D. L. Willis. 1985. Metabolism of ingested U and Ra. *Health Physics* 48(5):601-633.

3

Toxicology

This chapter presents information about the toxicology of uranium.[1] Studies of laboratory animals and other nonhuman systems are essential for understanding mechanisms of action, biologic plausibility, and possible health effects when experimental research in humans is not ethically or practically possible (Cohrssen and Covello, 1989; NRC, 1991). Such studies permit a potentially toxic agent to be introduced under conditions—such as dose, duration, and route of exposure—controlled by the researcher to probe health effects on many body systems. Nonhuman studies are also a valuable complement to human studies of genetic susceptibility. Although nonhuman studies often focus on one agent at a time, they enable investigation of chemical mixtures and their potential interactions more easily.

Research on health effects of toxic substances includes animal studies that characterize absorption, distribution, metabolism, and elimination. Animal studies may examine acute (short-term) exposures or chronic (long-term) exposures. Animal research may focus on the mechanism of action (how a toxicant exerts its deleterious effects at the cellular and molecular levels). Mechanism-of-action (or mechanistic) studies encompass an array of laboratory approaches with whole animals and with in vitro systems that use tissues or cells from humans or animals. Structure-activity relationships, in which the molecular structure and chemical and physical properties of a potential toxicant are compared with those of a known toxicant, are an important source of hypotheses about mechanism of action.

[1]Some sections of this chapter have been adapted from *Gulf War and Health, Volume 1: Depleted Uranium, Pyridostigmine Bromide, Sarin, Vaccines* (IOM, 2000).

In carrying out its charge, the committee used animal and other nonhuman studies in several ways, particularly to look for markers of health effects that might be important for humans. If animal studies showed absorption and deposition in specific tissues or organs, the committee looked especially closely for possible abnormalities at these sites in human studies, as it did for uranium deposition in bone and kidney. One of the problems with animal studies, however, is the difficulty of finding animal models to study symptoms that are related to uniquely human attributes, such as cognition, purposive behavior, and the perception of pain.

The toxic effects of uranium also have been reviewed in a recent National Research Council report, *Review of Toxicologic and Radiologic Risks to Military Personnel from Exposure to Depleted Uranium During and After Combat* (NRC, 2008), which assessed the US Army's "Capstone report" (USACHPPM, 2004) on toxicologic and radiologic risks to soldiers posed by exposure to depleted uranium.

This chapter begins with a summary of the findings presented in *Gulf War and Health, Volume 1: Depleted Uranium, Pyridostigmine Bromide, Sarin, Vaccines* (IOM, 2000), hereafter referred to as *Volume 1*. It next addresses experimental data from toxicokinetic (also called pharmacokinetic), animal, and in vitro studies published since *Volume 1*. The chapter ends with a discussion of how the committee applied the experimental data. Tables 3-1 to 3-10 are included at the end of this chapter.

SUMMARY OF PREVIOUS REPORT

Chapter 4 of *Volume 1* includes a review of studies of the toxicology of uranium. It covers toxicokinetics and animal and in vitro studies.

Uranium is both a heavy metal and a low-specific-activity radioactive element. Studies on the toxicity of uranium have examined both its chemical and its radiologic effects. The primary routes of human exposure to uranium are ingestion and inhalation; the effects of dermal exposure and embedded fragments have also been studied.

The amount of uranium that the body absorbs depends largely on the route of exposure and the solubility of the uranium compounds to which a person is exposed. Insoluble uranium compounds may remain in the pulmonary tissues, especially the pulmonary lymph nodes, for a long time and thus pose a localized radiologic hazard. As a general rule, uranium absorption from the intestinal tract is lower than that from the respiratory tract and results in lower doses per unit intake. Renal dysfunction and lung injury are the best-characterized consequences of exposure to uranium compounds. The chemical and radiologic properties of uranium could act cooperatively to cause tissue damage, so it cannot be assumed that excess cancers would be due solely to the radiologic effects of uranium or that organ damage is due exclusively to its heavy-metal properties.

TOXICOKINETICS

The toxicokinetics of a substance has to do with the routes and rates of absorption, distribution, metabolism, and excretion. Toxicokinetics can be used to determine the amount of a substance that reaches particular organs or cells and might therefore lead to a toxic effect. For a review on biokinetic models of exposure to different forms of uranium with emphasis on depleted uranium, the reader may consult Leggett (2006).

Absorption

Where uranium particles are deposited in the respiratory tract is the result of a combination of physical forces that govern particle behavior in an air stream and the anatomy of the respiratory tract (Gordon and Amdur, 1991). The site of deposition affects the degree of uranium absorption, the clearance mechanisms that are available to remove uranium particles, and the severity of the consequences of damage of tissue of the respiratory system.

Which of the various regions of the respiratory tract and lung (extrathoracic, tracheobronchial, and deep pulmonary or alveolar) inhaled uranium-dust particles are deposited in depends on the particles' aerodynamic diameter and inspiratory flow rate. An aerodynamic diameter that incorporates both the density and the diameter of particles and their aerodynamic drag is typically assigned to nonspherical particles. It represents a particle as having the diameter of a unit-density sphere that has the same terminal velocity as the particle, whatever its size, shape, or density (Gordon and Amdur, 1991). Larger particles are deposited in the tracheobronchial region; mucociliary action transports the particles to the pharynx, where they are swallowed. Smaller particles reach the terminal bronchioles and the alveoli. The International Commission on Radiological Protection has developed extensive models of the dosimetry of inhaled radioactive materials (ICRP, 2002).

At the alveolar level, the more soluble uranium compounds (categorized as type F for *f*ast dissolution) are taken up by the systemic circulation within days. The less soluble uranium compounds (type M for *m*edium dissolution) are likely to remain in the pulmonary tissue and associated lymph nodes for weeks. The relatively insoluble compounds (categorized as type S for *s*low dissolution) are least likely to enter the systemic circulation and may remain in the lung and tracheobronchial lymph nodes for several years (ATSDR, 1999). (See Table 3-1 for examples of uranium compounds of each type.) The lungs and the tracheobronchial lymph nodes are the two major sites of accumulation for type S uranium compounds (administered as uranium dioxide) in dogs, monkeys, and rats, accounting for greater than 90% of the total body burden of uranium after inhalation of the compounds (Leach et al., 1970).

Given their high density, most inhaled uranium-particle–containing dusts have an aerodynamic diameter that does not permit them to be carried to the peripheral part of the lungs (Berlin and Rudell, 1986; Morris et al., 1992). Estimates based on measurements in uranium-processing plants suggest that only 1-5% of uranium-particle–containing dusts will enter the lungs (Davies, 1961). The rest will deposit in the upper respiratory tract and eventually be swallowed and go through the gastrointestinal tract.

Inhalation studies of depleted-uranium particles in animals suggest that patterns of exposure may be important in the bioaccumulation of uranium (Monleau et al., 2006a,b). For example, repeated pre-exposure to insoluble depleted-uranium dioxide by inhalation has been shown to increase later uranium peroxide bioaccumulation in the kidneys and femurs and decrease it in the gastrointestinal tract and excreta concurrently with enhanced genotoxic effects in several of these tissues (Monleau et al., 2006b). No change in uranium peroxide bioaccumulation was found in the lungs.

Gastrointestinal absorption of uranium has been studied after single oral administration of soluble compounds to rats, swine, dogs, hamsters, and baboons (for review, see Thorne, 2003). The absorption of uranium in the gastrointestinal tract generally increases with increasing solubility of the compound, but only a small fraction of even the soluble uranium compounds is absorbed through the gastrointestinal epithelium. Gastrointestinal absorption was estimated at up to 5% and was highly variable between experiments. Uranium absorption occurs predominantly in the small intestine, and there is no absorption from the buccal cavity, stomach, or large intestine (Dublineau et al., 2005). The apparent uranium permeability measured with ex vivo techniques was similar in the various parts of the small intestine (Dublineau et al., 2005) and probably occurs through a transcellular pathway. The relationship between uranium speciation and gastrointestinal absorption was investigated in rats after ingestion of five samples of water contaminated with different forms of uranium. The average fractional absorption was about 0.4% for each of the samples, so it was concluded that the chemical form of uranium in the water did not influence its absorption into the systemic circulation (Frelon et al., 2005).

Peyer's patches, the aggregated structures of gut-associated lymphoid tissue, have specific electrophysiologic characteristics of ion conductance and secretory capacity (Brayden and Baird, 1994) and have recognized sites of transport of nanoparticles, microparticles, and macromolecules (Pappo and Ermak, 1989; Powell et al., 1996). A recent quantitative analysis of uranium deposition by inductively coupled plasma-mass spectrometry (ICP-MS) after chronic exposure of rats to depleted uranium during 3 or 9 months demonstrated preferential accumulation of uranium in Peyer's patches compared with epithelium (Dublineau et al., 2006a). However, the apparent uranium permeability of the rat intestine was higher (by a factor of 10) in the mucosa than in Peyer's patches, and this suggests that the small intestinal epithelium was the preferential pathway for the transmucosal passage of uranium.

As described in *Volume 1*, skin absorption is an effective route for the entry of soluble uranium compounds into the systemic circulation. Substantial diffusion of soluble uranium through human or mammalian intact skin has been described (de Rey et al., 1983; Lopez et al., 2000; Tymen et al., 2000; Petitot et al., 2004).

Absorption of uranium via deep wounds has also been shown to depend on the solubility of the metal. Pellmar et al. (1999a) assessed distribution of uranium that was implanted in the gastrocnemius (lower leg) muscle of rats in the form of depleted-uranium pellets. A dose-dependent increase in uranium concentration was noted 1 day after implantation, reaching 82.0 ± 9.7 and 31.3 ± 6.5 ng of uranium per gram in the kidneys and tibias, respectively, of high-dose animals. Those concentrations were about 58 and 26 times higher than seen in the same tissues of the control group.

A recent study evaluated the influence of wounds on the short-term distribution and excretion of uranium in rats (Petitot et al., 2007). The authors reported substantial uptake of a uranyl nitrate solution through intact rat skin within the first 6 hours of exposure. Skin excoriation increased percutaneous absorption of uranyl nitrate, and this suggested that percutaneous diffusion of uranyl nitrate depends heavily on compromised skin-barrier integrity (Petitot et al., 2007). Similar studies with other forms of uranium were not found. However, in vitro studies corroborated greater diffusion of uranium through excoriated skin than intact skin; substantial uptake of uranium through excoriated skin occurred as early as 30 minutes after exposure (Petitot et al., 2004).

Transport and Biotransformation

Once absorbed, uranium forms soluble complexes with bicarbonate, citrate, or proteins in the plasma (Dounce, 1949; Stevens et al., 1980; Cooper et al., 1982). Little is known about the cellular and molecular mechanisms underlying the uptake of uranium in tissues. In the kidneys, a cytotoxic fraction of uranium was found to be a phosphate complex of uranyl whose uptake is mediated by a sodium-dependent phosphate cotransporter system (Muller et al., 2006). No other information on the mechanisms of transport could be found. The role of nonspecific metal transporters, such as divalent metal transporter-1, in the transport of uranium has yet to be defined.

Distribution

The percentages of uranium absorbed into blood, transferred to tissues, and excreted in urine are independent of the deposition of soluble uranium compounds, such as uranium peroxide or uranium tetrafluoride dust, in the lungs (Houpert et al., 1999). The ratio of K to (K + U), where K equals the percentage of uranium retained in the kidneys and U equals the percentage excreted in urine 24 hours after instillation, may be used to characterize kidney clearance of uranium. The ratio was constant when the concentration of uranium, in the form of the two compounds mentioned, in the kidneys increased from 0.02 to 12.5 μg/g.

Inhaled uranium accumulates readily in the central nervous system (Monleau et al., 2005). Repeated exposure (4 days/week for 3 weeks) to depleted-uranium dioxide at a high air concentration (197 mg/m^3) has been associated with uranium accumulation in the following rank order: olfactory bulb > hippocampus > frontal cortex > cerebellum. Accumulation in brain regions appears to be route-dependent: injection of uranium results in homogeneous distribution in various brain regions, whereas inhalation and ingestion result in heterogeneous and specific accumulation (Houpert et al., 2007c). Those differences were thought to reflect differential mechanisms of delivery of uranium to the brain; however, the nature of the transporters remains unknown. Lemercier and colleagues (2003) demonstrated transfer of uranium across the blood-brain barrier in an in situ rat brain perfusion study in which brain uranium was measured with ICP-MS.

In chronic exposure to uranium, tissue deposition does not appear linear (Paquet et al., 2006). For example, Tracy et al. (1992) showed that uranium concentrations in rat femurs were 6.3 times higher after 28 days of ingestion of uranyl nitrate in drinking water than after 91 days of ingestion. In the case for the kidneys, uranium concentration increased by a factor of 1.8 from 28 to 91 days. Similarly, in Sprague Dawley rats that had implanted depleted-uranium pellets, Pellmar et al. (1999a) showed that uranium concentrations in kidneys peaked 6 months after exposure began and then decreased by a factor of 1.4-1.6 until 18 months of exposure. That pattern of accumulation was also observed in urine, in which a peak uranium concentration was noted at 12 months of exposure. Similarly, rats exposed to uranyl nitrate via drinking water during their entire adult life at a constant concentration of 40 mg/L showed fluctuations in uranium concentration in almost every tissue (Paquet et al., 2006); this suggested that accumulation of uranium in tissues may vary over the course of chronic exposure. Thus, chronic exposure may be associated with physiologic phenomena that modify the pharmacokinetics of uranium over time.

Excretion and Retention

No new studies were identified on systemic clearance or excretion and retention of uranium after inhalation or oral exposure. As described in *Volume 1*, Pellmar and colleagues (1999a) reported that bone and kidneys were the primary reservoirs of uranium that had dissolved from embedded depleted-uranium fragments. Dissolved uranium also localized in various nuclei of the brain, lymph nodes, testes, and spleen, and low serum concentrations of uranium were noted at all times of measurement whereas the size of the pellets diminished with time (Pellmar et al., 1999a; Fitsanakis et al., 2006). Adult male and female rats with surgically implanted depleted uranium were shown to excrete uranium in urine in a dose-dependent manner, but tissue concentrations of uranium were not reported (Arfsten et al., 2005). In mice, 60 days after implantation of depleted uranium, the highest concentration of uranium was noted in the kidneys, but bone marrow, hind limbs, and spleen also had substantial increases in uranium (Miller et al., 2005). Those studies establish that depleted uranium from implanted fragments will readily distribute to various tissues over the life span of an animal. Additional details from multiple studies on the accumulation of uranium in the brain are presented below in the section on "Nonmalignant Neurologic Effects."

TOXICITY STUDIES

This section reviews key animal and in vitro studies of the toxic effects of uranium published since *Volume 1*. Studies of cancer and noncancer health end points have used inhalation and oral and dermal exposure. There are also studies of the effects of injected uranium and embedded depleted-uranium fragments. Kathren and Burklin (2008) have proposed a median lethal dose for acute oral intake of uranium in humans of 5.0 g and for acute inhalation of soluble uranium compounds of 1.0 g.

Carcinogenic Effects

Four studies of the carcinogenic effects of uranium were described in *Volume 1*; two reported positive findings (Leach et al. 1973; Filippova et al., 1978). Since the publication of *Volume 1*, three studies were found that examined cancer in depleted-uranium–exposed animals. The experimental details are presented in Table 3-2.

In the first study, rats received thigh-muscle implants of depleted-uranium pellets or fragments and were held for their life span (Hahn et al., 2002). At death, necropsies and histopathologic examinations were conducted. A statistically significant increase in the incidence of soft-tissue sarcoma at the implantation site was reported in rats that received 5.0 × 5.0-mm squares of depleted uranium and a slight increase in rats that received 2.5 × 2.5-mm squares. No tumors were observed in the rats that received 2.0 × 1.0-mm depleted-uranium pellets.

The second study used an in vivo leukemogenesis mouse model (Miller et al., 2005). Mice received implants two to eight depleted-uranium pellets in the gastrocnemius muscle, and 60 days later received intravenous injections of murine multicolony-stimulation-factor–dependent hematopoietic cells. Leukemia developed in 68-75% of the mice that received depleted-uranium implants and 12% of the mice that had no implants.

In the third study, Mitchel et al. (1999) exposed rats by nose-only inhalation 4.2 hours/day 5 days per week for 65 weeks to natural uranium-ore dust aerosol (44% uranium) in the absence of substantial radon content at 50 or 19 mg/m^3. Lung uranium burdens, determined at the time of death, decreased exponentially after cessation of exposure independently of the initial burden. The frequency of primary malignant lung tumors was dose-dependent: 0.016, 0.175, and 0.328 in the control, low-aerosol, and high-aerosol groups, respectively. The groups were indistinguishable with respect to tumor latency. The average radiation doses received at the low and high dust-aerosol concentrations were 0.87 and 1.64 Gy, respectively, and resulted in an average risk of malignant lung tumors of about 0.20 tumor per animal per gray in both exposure groups. The tumor frequency was not directly proportional to the dose; but when malignant lung-tumor frequency was calculated as a function of dose rate (measured on the basis of lung burden at the end of dust inhalation), a direct linear relationship was noted and suggested that radiation dose rate may be a more important determinant of lung-cancer risk than absolute chemical exposure. The authors concluded that chronic inhalation of natural uranium-ore dust alone in rats creates a risk of primary malignant lung-tumor formation. The study also provided details on urinalysis, which was carried out once per month throughout the animals' lifetime. The results demonstrated a constant urinary concentration of uranium throughout weeks 13-55 of exposure, with average concentrations of 0.117 and 0.274 mg/L in the low-uranium and high-uranium groups, respectively. Mitchel et al. (1999) also assessed the effects of inhaled uranium on the incidence of nonmalignant lung tumors in rats. They exposed rats to one of two concentrations of natural uranium-ore dust aerosol by nose-only inhalation for 4.2 hours/day 5 days per week for 65 weeks. The proportion of animals with nonmalignant lung tumors was 0.016, 0.135, and 0.131 in the control, low-aerosol, and high-aerosol groups, respectively.

Genotoxic Effects

Volume 1 describes two genotoxicity studies (Miller et al., 1998a,b), both of which reported positive findings. A number of studies have been conducted since *Volume 1* on genotoxic effects of depleted uranium in humans. In a 10-year postwar followup assessment, 13 Gulf War veterans with high concentrations of depleted uranium from embedded fragments had a statistically significantly higher incidence of chromosomal aberrations in their peripheral blood lymphocytes than 26 in the low-exposure group (McDiarmid et al., 2004). However, no significant difference in chromosomal aberrations was reported between high-exposure and low-exposure groups in the 8-, 12-, and 14-year postwar followup assessments (McDiarmid et al., 2001; McDiarmid, June 28, 2007, presentation to the committee). Hypoxanthine-guanine phosphoribosyl transferase mutation frequencies were nonsignificantly higher in the high-exposure group than in the low-exposure group (McDiarmid, June 28, 2007, presentation to the committee). An increase in chromosomal aberrations in peripheral blood lymphocytes was reported in a cohort of 69 people in southern Serbia and Montenegro who were exposed to depleted uranium during air strikes in 1999 (Milacic et al., 2004). Statistically significant increases in micronuclei frequencies in peripheral blood lymphocytes were found in 30 people who lived near Sarajevo (Krunic et al., 2005). Urinalysis to determine body burden of uranium was not conducted on that cohort, but water from two local wells contained traces of depleted uranium. As detailed in Table 3-3, uranium-induced genotoxicity was demonstrated in a number of studies of rats; human, hamster, and rat cells; and calf thymus DNA.

Respiratory Effects

Six toxicologic studies of respiratory effects in several animal species are described in *Volume 1*; their results were inconsistent. Several in vitro studies of the effects of uranium on lung epithelial cells and macrophages have been published recently; details are presented in Table 3-4.

In rat lung epithelial cells, treatment with uranyl (VI) acetate was associated with increased oxidative stress and decreased cell proliferation (Periyakaruppan et al., 2007). The authors attributed the decrease in cell proliferation to loss of total cellular redox potential due mainly to depletion of the tripeptide glutathione and superoxide dismutase.

In light of uranium's ability to induce pulmonary fibrosis, which is often associated with inflammation, Gazin et al. (2004) evaluated the effects of uranium on cytokine secretion—tumor-necrosis factor-alpha (TNF-alpha), interleukin-1beta (IL-1beta), and IL-10—in alveolar macrophages. TNF-alpha secretion was increased by exposure to uranium but not by exposure to the metallic element gadolinium. Uranium-treated and control cells were indistinguishable with respect to IL-1beta and IL-10 secretions.

In another study, a 48-hour exposure of a human type II epithelial cell line (A549) to 0.5 mM uranyl bicarbonate solution triggered differential expression of cytokeratin 8 (CK8) and CK8 fragments; this suggested dysfunction of the ubiquitin-proteasome system or a regulator pathway involving CK ubiquitinylation (Malard et al., 2005).

Renal Effects

It is well established that uranium causes low-level metallotoxic effects on the renal system in animals (IOM, 2000). In general, renal injury occurs within days of exposure and is manifested as a change in the proximal convoluted tubules, which results in increased urinary enzyme excretion (excretion of alkaline phosphatase, lactate dehydrogenase, and leucine aminopeptidase). Hyaline casts (casts containing necrotic cells shed from the tubular epithelium) are present at all levels of the tubular system (Berlin and Rudell, 1986). Glomerular changes occur in parallel with tubular damage, principally in the basement membranes of glomerular capillaries. The corresponding functional changes in the kidney are proteinuria, impairment of *p*-aminohippurate clearance, increase in clearance of amino acids and glucose, and decrease in sodium reabsorption. After severe damage, renal inulin and creatinine clearance decreases (Stopps and Todd, 1982). Generally, if the uranium dose is sublethal, regeneration of the damaged epithelium commences within 2-3 days of the end of exposure (Stopps and Todd, 1982; Berlin and Rudell, 1986; Gilman et al., 1998a; ATSDR, 1999). Renal injury was not observed in dogs and monkeys exposed for 5 years to inhaled dust that contained insoluble uranium dioxide at a uranium concentration of 5 mg/m^3 (Leach et al., 1970). After a 91-day exposure to uranyl nitrate hexahydrate in drinking water at 0.96, 4.8, 24, 120, or 600 mg/L, histopathologic lesions were observed in the kidneys of male and female New Zealand white rabbits in all groups, including the lowest-exposure groups (Gilman et al., 1998b). Pathologic changes included lesions of tubular epithelial cells (apical nuclear displacement and vesiculation, cytoplasmic vacuolation, and dilation), glomeruli (capsular sclerosis), and renal interstitium (reticulin sclerosis and lymphoid cuffing). Studies of dermal and ocular absorption of uranium trioxide in rabbits indicated that uranium was sufficiently well absorbed to cause renal damage and even death from renal failure (Voegtlin and Hodge, 1949).

Several mechanisms may account for uranium-induced renal damage. A mechanism involving bicarbonate activity in the kidney has been postulated. Uranium combines with bicarbonate, citrate, or plasma proteins in blood. At low pH, the bicarbonate-uranyl and citrate-uranyl complexes split (Bassett et al., 1948), and the resulting uranyl ion may combine with proteins on the tubular wall and cause renal damage. A second possibility is that uranium compounds inhibit mitochondrial oxidative phosphorylation and sodium-dependent and sodium-independent adenosine triphosphate (ATP) use in renal tubules (Brady et al., 1989).

Few animal and in vitro studies of renal effects of depleted uranium have been conducted since *Volume 1* (see Table 3-5). In one study, concentrations of *N*-acetyl-β-D-glucosaminidase and creatinine in rats given depleted uranium nitrate by a single intramuscular injection peaked 3 days later, and there was a high correlation between injected dose and those concentrations (Fukuda et al., 2006). Depleted-uranium concentrations in urine decreased rapidly for the first 3 days after exposure. Another study reported a decrease in glucose transport in brush-border membrane vesicles of rats given depleted uranium in the form of uranyl acetate (Goldman et al., 2006). Donnadieu-Claraz et al. (2007) exposed rats to uranium nitrate at 40 mg/L of water for up to 18 months and observed that the proximal tubular cells had more vesicles with dense granular inclusions. The granules were found to be iron oxides; uranium was not associated with them. The authors suggested that the mechanisms of iron homeostasis in the kidneys could be affected by chronic uranium exposure.

Neurologic Effects

The earlier animal studies considered in *Volume 1* indicated that uranium crosses the blood-brain barrier and deposits within the brain parenchyma. In this volume, several relevant in vitro studies will be discussed in addition to animal studies on neurological effects published since *Volume 1*. Study details can be found in Table 3-6.

In Vitro Models to Assess Neurologic Effects

Lemercier and colleagues (2003) demonstrated transfer of uranium across the blood-brain barrier in an in situ rat brain perfusion study in which brain uranium was measured with ICP-MS. They found that substantial uranium accumulated in the brain and that the transport was efficient and rapid, with uranium localized to the brain parenchyma as early as 2 minutes after the initial perfusion. The nature of the uranium transporter is unknown. No functional end points were assessed in the study.

One of the earliest studies of the specific effects of uranium on functional end points in the nervous system focused on the presynaptic action in phrenic nerve preparations from mice. This in vitro study demonstrated that uranyl nitrate at very high concentrations (0.2-0.8 mM) facilitated the release of acetylcholine from the nerve terminals and potentiated muscle contraction (Lin et al., 1988).

The cytotoxicity of depleted uranium was investigated in an in vitro model of the blood-brain barrier with rat brain endothelial cells (RBE4 cells). The cells were derived from rat brain microvascular endothelial cells immortalized with the plasmid pE1A-neo that contained the E1A region of adenovirus 2 and a neomycin-resistance gene. Cytotoxicity was evaluated with assays for cell-volume increase, heat-shock protein 90 expression, 3-[4,5-dimethylthiazol-2-yl]-2,5-diphenyltetrazolium bromide (MTT) reduction, and lactate dehydrogenase (LDH) activity. The results of the assays showed that uptake of the triuranium octaoxide uranyl chloride form of depleted uranium into RBE4 cells is efficient, but no overt cytotoxicity was detected by common biomarkers (Dobson et al., 2006).

In vitro studies in rat cortical neuron cultures exposed to uranyl acetate showed little cytotoxicity at concentrations below 100 μM (Jiang et al., 2007). There were no statistically significant changes in F2-isoprostanes, biomarkers of oxidative stress, or thiol metabolites. The lack of cytotoxicity was corroborated with the MTT-reduction and LDH-activity assays and the finding of only minimal changes in total adenosine nucleotides. Additional studies in *Caenorhabditis elegans* using green fluorescent protein reporter worm strains corroborated the primary neuron culture observations, showing no statistically significant neuronal degeneration after uranium exposure (Jiang et al., 2007).

Although in vitro models may not recapitulate human health and disease fully, those focused studies indicate that cultured neurons can tolerate high concentrations of uranyl acetate without important oxidative injury or death (Dobson et al., 2006; Jiang et al., 2007). The studies have examined neuronal lethality after a relatively acute exposure, but impairment of neuronal function may occur in the absence of acute lethality, and effects might arise after longer exposure. Other types of brain cells, such as astrocytes, may be more sensitive to uranium, but this seems unlikely, given their greater redox potential and sensitivity to oxidative stressors compared with neurons and endothelial cells used in the test systems discussed above (Dringen and Hirrlinger, 2003).

In Vivo Studies to Assess Neurologic Effects

Abou-Donia et al. (2002) investigated the effects of uranyl acetate on sensorimotor behavior, generation of nitric oxide, and the central cholinergic system of rats. Intramuscular injection of uranyl acetate at 0.1 and 1.0 mg/kg for 7 days daily was followed by a 30-day observation period. On cessation of treatment, the sensorimotor functions of the animals were evaluated with a battery of tests that included measurements of postural reflexes, limb placing, orientation to vibrissa touch, grip time, beam walking, and inclined-plane performance. Treatment with uranyl acetate was associated with dose-dependent deficits in inclined-plane performance, beam-walk score, and beam-walk time. Changes in nitric oxide were inconsistent, increasing in cortex and midbrain and decreasing in brainstem and cerebellum at both doses. Acetylcholinesterase (AChE) activity in the cortex, but not other brain regions, of the animals given the high dosage was statistically significantly increased. Ligand-binding densities for the M2 muscarinic receptor did not show any change.

The acute and chronic consequences of exposure to depleted-uranium fragments in animal models have been addressed by a number of groups (Gilman et al., 1998a,c; Pellmar et al., 1999a,b; Barber et al., 2005; Houpert et al., 2005; Lestaevel et al., 2005b; Monleau et al., 2005; Fitsanakis et al., 2006). *Volume 1* described studies conducted by Pellmar et al. (1999a) to establish an animal model providing insight into the injuries sustained by Gulf War veterans from embedded depleted-uranium fragments and to evaluate the biologic effects of intramuscularly embedded depleted-uranium fragments. Pellmar (1999a) suggested that uranium can accumulate in the rat central nervous system. An earlier study by Pellmar (1997) suggested that retained depleted-uranium fragments in the leg muscle are associated with increased brain uranium concentrations.

Studies by Fitsanakis et al. (2006) corroborated the earlier observations by Pellmar et al. (1999a). A similar model of surgical intramuscular implantation randomly assigned rats to five groups: nonsurgical control (NS control); no depleted-uranium pellets and 20 tantalum (Ta) pellets (sham); four depleted-uranium pellets and 16 Ta pellets (low); 10 depleted-uranium pellets and 10 Ta pellets (medium); and 20 depleted-uranium pellets and no Ta pellets (high). Uranium content was measured in digested samples as ^{238}U with high-resolution ICP-MS. Three months after implantation, depleted uranium had accumulated substantially in the high group in all brain regions except the hippocampus. By 6 months, however, substantial accumulation was measured only in the cortex, midbrain, and cerebellum in the medium and high groups.

Barber et al. (2005) studied rats treated with single intraperitoneal injections of uranium as uranyl acetate at 1 mg/kg to determine the temporal and regional distribution of depleted uranium in the brain, as measured with ICP-MS. In agreement with studies of Pellmar et al. (1999a,b) and Fitsanakis et al. (2006), uranium readily accumulated in the brain. At 24 hours after exposure, it was concentrated in the hippocampus, striatum, cerebellum, and cortex. Prior exposure to stress (five daily episodes of forced swimming) statistically significantly reduced hippocampal and cerebellar uranium then and tended to reduce uranium in all brain regions 7 days after exposure. The authors did not discuss the reason for the increased clearance or allude to the effects of uranium on functional end points in this acute-exposure model.

Functional consequences of exposure of rats to depleted uranium in drinking water were addressed by Briner and Murray (2005), who found that depleted-uranium exposure was associated with short-term and long-term differences in brain lipid oxidation and open-field behavior. After 2 weeks of exposure to depleted uranium, brain lipid oxidation, measured with the thiobarbituric acid assay, was increased and correlated with increases in line-crossing and rearing behavior. Although the open-field behavior differences were sustained after 6 months of exposure, brain lipid oxidation did not correlate with the behavioral changes at that time. Male rats appeared to be more sensitive than female rats to the behavioral effects of depleted uranium.

Monleau et al. (2005) exposed rats to depleted-uranium dioxide by inhalation (depleted uranium at 197 mg/m^3) for 30 minutes a day 4 days a week for 3 weeks. They observed greater spontaneous locomotion activity in treated rats than in controls on the first day after the end of the exposure period and attenuated spatial working memory on the sixth day after exposure. Houpert et al. (2007b) noted decrements in performance on open-field, Y-maze, and elevated plus-maze tests in 2-, 5-, and 9-month-old rats exposed to enriched uranium nitrate (40 mg/L) during gestation and lactation. The effects were consistent with decreased spatial working memory and mimicked the effects in adult rats (Monleau et al., 2005). Additional effects included delayed hyperactivity in the uranium-exposed rats.

Exposure to 4% enriched uranium or depleted uranium in drinking water for 1.5 months resulted in substantial accumulation of uranium in the hippocampus, hypothalamus, and striatum of rats, and the uranium concentrations were consistently 1.5-2 times higher in the hippocampus, hypothalamus, and adrenal of rats exposed to enriched uranium than of rats exposed to depleted uranium or control rats (Houpert et al., 2005). Increased brain uranium concentrations were associated with a statistically significant increase in the amount of paradoxic sleep, a reduction in spatial working memory capacities, and an increase in anxiety. Sleep-wake cycle disturbances characterized by an increase in rapid eye movement (REM) sleep and theta-band power during the light period were also noted in the rats as early as 30 days after exposure to depleted uranium in drinking water at 40 mg/L (Lestaevel et al., 2005a). A large variety of neuronal structures and substances, including neurotransmitters and peptides, are known to modulate REM sleep; the underlying mechanisms for these changes remain unknown.

Using a similar exposure paradigm (1-month exposure to depleted uranium at 40 mg/L in drinking water), the same group of investigators examined dopamine and serotonin brain metabolism in the rat (Houpert et al., 2004). No statistically significant differences were found in dopamine, serotonin, and their catabolite levels in the striatum, hippocampus, cerebral cortex, thalamus, or cerebellum between depleted uranium-exposed and control rats. Thus, it appears that depleted uranium-induced changes in the sleep-wake cycle are not mediated by dopamine and serotonin. Other neurotransmitters might be involved, or, given the role of the hypothalamic-pituitary axis in sleep regulation, these effects may be modulated by glucocorticoids. Bussy et al. (2006) investigated effects of exposure to depleted uranium (as uranyl nitrate) in drinking water at 40 mg/L for up to 9 months on dopaminergic and serotoninergic metabolism in rats and found subtle and transient perturbations of monoamine concentrations and AChE activity in discrete brain areas.

Biologic Plausibility

Collectively, the results of those studies indicate that depleted uranium is a toxicant that can cross the blood-brain barrier and might produce some acute and prolonged behavioral changes. The mechanisms associated with the effects are difficult to reconcile, given the differences in exposure models and apparent contradictions in the results: some studies established a treatment effect and others failed to. Although at high concentrations different forms of uranium might be associated with some subtle neurologic dysfunction, the significance of the observations is unknown.

Gastrointestinal Effects

Absorption of ingested uranium occurs mainly in the small intestine (ICRP, 1979). The chemical form of uranium in ingested water does not appear to influence absorption from the gastrointestinal tract in rats (Frelon et al., 2005). No animal or in vitro studies on the effects of uranium-induced pathology in the gastrointestinal tract could be found since those described in *Volume 1*. As summarized in *Volume 1*, studies have suggested that up to 2-year exposure of various animal species to high doses of uranium nitrate is not associated with gastrointestinal effects. Studies that used other forms of uranium could not be found.

Hepatotoxicity

Uranium-induced hepatotoxicity has not been a prominent finding in most animal studies (ATSDR, 1999). A few studies of hepatotoxicity have been published since *Volume 1*; the experimental details are summarized in Table 3-7.

In adult male zebrafish exposed to exceedingly high depleted-uranium concentrations in water, increased hepatic oxidative stress was observed (Barillet et al., 2007) in association with decreases in superoxide dismutase and catalase activity and in total glutathione content.

Exposure of rats to high concentrations of depleted uranium (11.5 mg/kg) by subcutaneous administration altered hepatic metabolism of bile acids and xenobiotics through modulation of cytochrome P450 (CYP) isoenzymes (Gueguen et al., 2006).

Subtle effects have been reported in rats exposed for 9 months to depleted uranium in drinking water and then treated with acetaminophen (*N*-acetyl-*p*-aminophenol, APAP) (Gueguen et al., 2007). Plasma concentration of APAP was higher and hepatic CYP activities were lower in the group exposed to depleted uranium than in controls. Furthermore, APAP treatment in the depleted-uranium rats was associated with a more rapid increase in plasma alanine aminotransferase and aspartate aminotransferase.

Using a similar exposure paradigm, the same group (Souidi et al., 2005) addressed the effects of 9 months of exposure to depleted uranium in drinking water on drug-metabolizing enzymes. Hepatic CYP3A1 and CYP3A2 mRNA expression was statistically significantly higher in rats exposed to depleted uranium than in controls, but CYP1A1 mRNA expression was not different. Nuclear pregnane X receptor (PXR) mRNA increased, but 9-*cis*-retinoic acid receptor mRNA was unchanged in the course of the study. Hepatic activity of CYP2C, CYP3A, CYP2A, or CYP2B remained indistinguishable in the depleted-uranium group compared with the controls. The authors hypothesized that uranium may affect the expression of drug-metabolizing CYP enzymes through the PXR and constitutive androstane receptor and thus potentially interfere with the metabolism of xenobiotics.

Reproductive and Developmental Effects

Volume 1 reported that modest testing of mice orally exposed to uranium did not establish association with reproductive or developmental problems but that no animal studies had used dermal or inhalation exposure. Developmental effects of depleted uranium are of particular interest because developing animals are known to be sensitive to the effects of metals (Briner, 2007). There is some recent evidence that exposure of animals to uranium compounds, including depleted uranium, during development can lead to a variety of adverse effects (recently reviewed in McClain and Miller, 2007, and Briner, 2007). Several studies published since *Volume 1* are summarized in Table 3-8.

Subcutaneous injection of uranyl acetate dehydrate at 0.415 or 0.830 mg/kg per day in rats on gestation days 6-15 led to reduced fetal body weight and an increase in the total number of skeletally affected fetuses (Albina et al., 2003).

Arfsten et al. (2005, 2006) reported some results of a large multigeneration reproductive study in which 12 pellets composed of depleted uranium, tantalum, or steel in various combinations were implanted in the gastrocnemius muscles of nine groups of 21 male and 21 female rats. The parental generation was mated 30 or 120 days after implantation. The depleted uranium did not adversely affect reproduction in the parental generation with respect to effects on male reproductive success, sperm concentration, and sperm velocity (Arfsten et al., 2006), and no increased incidence of developmental effects (birth weight, survival, litter size, gross physical abnormalities, neurodevelopmental effects, and immune function effects) was identified in the offspring of the mating at 30 days after implantation (Arfsten et al., 2005).

Briner and Byrd (2000) reported that offspring of female mice exposed to uranium acetate at up to 75 mg/L in drinking water for 2 weeks before mating and during gestation and lactation showed more rapid development on some behavioral tests whereas their ratio of brain weight to body weight was statistically significantly lower than in controls. However, the quickened development of uranium-exposed offspring may adversely affect the development of neural systems (Briner, 2007). In contrast, decrements in performance on open-field, Y-maze, and elevated plus-maze tests were noted in 2-, 5-, and 9-month-old rats whose mothers had been exposed to enriched uranium nitrate at 40 mg/L in drinking water for 3 months before mating and during gestation and lactation (Houpert et al., 2007b).

Raymond-Whish et al. (2007) recently addressed the effects of uranium hexahydrate in intact, ovariectomized, or pregnant mice at 0.5 μg/L (0.001 μM) to 28 mg/L (120 μM) in drinking water. The study was conducted to assess whether uranium added to the drinking water causes responses in the female mouse reproductive tract that might be accounted for by inherent estrogenic effects of uranium. Increased uterine weight and uterine luminal epithelial cell growth, selective reduction in ovarian primary follicles but increase in growing follicles, accelerated vaginal opening, and persistent presence of cornified vaginal cells were noted in mice that drank uranium, and these effects could be attenuated by coadministration of an antiestrogenic compound (ICI 182,780). Transplacental exposure to uranium was associated with fewer primordial follicles in developing pup ovaries. The estrogenic responses noted above occurred at or below the US Environmental Protection Agency safe drinking-water concentration of 30 μg/L (0.126 mM) (EPA, 2006). Uranium concentrations in blood and urine were not reported in the study, so it is difficult to extrapolate the results to humans.

Immune System Effects

Immunologic effects were not addressed in *Volume 1*, but several experimental studies of toxic effects of uranium on the immune system have been published. They are presented below and in Table 3-9.

In vitro studies have investigated the effects of depleted uranium on cells of the immune system. One addressed the effects of depleted uranium (as uranyl nitrate) on viability and immune function and on cytokine gene expression in murine peritoneal macrophages and splenic CD4+ T cells (Wan et al., 2006). Depleted uranium was shown to affect signal-transduction pathways (c-jun and NF-kappa Bp65), neurotrophic factors (Mdk), chemokine and chemokine receptors (TECK/CCL25), and IL-10 and IL-5 concentrations (Wan et al., 2006). Depleted uranium also led to apoptosis in macrophages and CD4+ T cells (Kalinich et al., 2002; Wan et al., 2006). However, those effects were manifested at uranium concentrations exceeding 100 μM, so the relevance of the findings to in vivo scenarios is uncertain.

Increased spleen and lymphoid tissue uranium concentrations have been noted after chronic inhalation of natural uranium in adult rats (Leach et al., 1970, 1973), and depleted uranium has also been shown to accumulate in rat immune organs (Pellmar et al., 1999a), but there is little and contradictory information on the in vivo effects of uranium in general and depleted uranium in particular on immune function in animals.

Results of the multigeneration reproductive study discussed above show that in offspring of adult rats that had implanted depleted-uranium pellets there are no statistically significant effects on immune function (Arfsten et al., 2005). Mean thymus and spleen weights and the mean total number of thymocytes per spleen were indistinguishable from the values in a control group.

No effects on the intestinal localization and density of neutrophils, helper T lymphocytes, and cytotoxic T lymphocytes were observed 1-3 days after gavage treatment of rats with depleted uranium at 204 mg/kg despite modulation of cytokine (interferon-gamma [IFN-gamma]) and chemokine (MCP-1) expression (Dublineau et al., 2006b). In contrast, chronic exposure to depleted uranium in drinking water at 40 mg/L for 3, 6, or 9 months was associated with decreased intestinal mast-cell number, increased IL-1beta and IL-10 concentrations, decreased mRNA CCL-2 concentrations, decreased intestinal macrophage density, and increased numbers of neutrophils (Dublineau et al., 2007). The same group (Dublineau et al., 2006a) found no change in cytokine expression patterns (IL-10, transforming growth factor-beta, IFN-gamma, TNF-alpha, and monocyte chemoattractant protein-1) in the Peyer's patches from intestines of rats similarly exposed to depleted uranium in drinking water.

Increases in inflammatory cytokine expression and production of hydroperoxides in lung tissue from rats indicated that the genotoxic damage may be a result of the inflammatory processes and oxidative stress (Monleau et al., 2006c).

Cardiovascular Effects

No additional animal or in vitro studies of the cardiovascular effects of uranium have been identified since *Volume 1*. As summarized in *Volume 1*, studies of high doses of uranium in several animal models suggested that the cardiovascular system is not a sensitive target for this metal.

Dermal Effects

No additional animal or in vitro studies of dermal effects of uranium have been identified since *Volume 1*. In several studies reviewed in *Volume 1,* dermal application of uranium compounds was associated with mild skin irritation, severe dermal ulcers, or superficial coagulation necrosis and inflammation of the epidermis in rabbits and swollen and vacuolated epidermal cells and damage to hair follicles and sebaceous glands in rats. Inhalation and oral exposures to uranium compounds have not led to dermal effects in animals.

Ocular Effects

No additional animal or in vitro studies of ocular effects of uranium have been identified since *Volume 1*. Only two studies were reviewed in *Volume 1*; they reported encrusted eyes and conjunctivitis in animals after direct contact of the eye with uranium aerosol or vapor.

Musculoskeletal Effects

The effects of inhaled uranium on the musculoskeletal system of animals have not been examined. Studies with intravenously administered uranium in the late 1940s were the first to establish uranium's high affinity for bone. About 20-30% of intravenously administered uranium could be found in bone within 2.5 hours of administration, and 90% of the uranium retained in the body 40 days after injection was in bone (Neuman et al., 1948a). Greater amounts of uranium were incorporated in bones of young rats and calcium-deficient mature rats than normal mature rats (Neuman et al., 1948b). Uranium was specifically incorporated in areas of active calcification; the areas of uranium deposition became refractory to resorption as new calcification covered them (Neuman and Neuman, 1948).

There were no histopathologic findings in rat or rabbit muscles after exposure to orally administered uranyl nitrate in drinking water at uranium concentrations up to 40 mg/kg per day for 28 day or up to 53 mg/kg per day for 91 days in Sprague-Dawley rats or up to 53 mg/kg per day for 91 days in rabbits (Gilman et al., 1998a,b). However, acute uranium intoxication in suckling rats given [^{238}U]uranyl nitrate at a uranium concentration of 2 mg/kg of body weight intraperitoneally has been shown to inhibit bone formation and mandibular growth; this effect is believed to be due to the direct action of uranium on bone-forming cells or their precursors (Guglielmotti et al., 1985; Ubios et al., 1998). Inhibition of bone formation by uranium has been shown in endochondral ossification (Guglielmotti et al., 1984), alveolar bone healing (Guglielmotti et al., 1985, 1987), and alveolar bone modeling and remodeling (Ubios et al., 1990). Bone biochemical markers—such as osteocalcin, tartrate-resistance acid phosphatase, pyridinoline, and rat parathyroid hormone—were increased 28 days after rats received single intramuscular injections of depleted uranium and indicated bone damage (Fukuda et al., 2006). That study and other recent studies of musculoskeletal effects of uranium exposure are detailed in Table 3-10.

Pujadas Bigi and Ubios (2007) noted that after a single injection of uranyl nitrate in 1-day-old rats there was statistically significant but transitory inhibition of tooth eruption, dental development, and mandibular growth retardation; the delay in dental growth was attributed to damage to the odontoblast and cementoblast cell lineage.

Other studies in which rats were orally exposed to depleted uranium found statistically significant decreases in expression of CYP27A1, CYP2R1, CYP27B1, and CYP24A1 enzymes involved in vitamin D metabolism and two vitamin D(3)-target genes (ECaC1 and CaBP-D9K) (Tissandie et al., 2006, 2007). Although depleted-uranium–induced changes in the concentrations of the active form of vitamin D and its receptor expression could potentially be associated with the modulation of the expression of vitamin D-target genes and calcium homeostasis and thus effects on bone deposition and remodeling, it is difficult to ascribe physiologic significance to these findings, given the high doses of depleted uranium to which the animals were exposed.

Hematologic Effects

No additional animal or in vitro studies of hematologic effects of uranium have been identified since *Volume 1*. Results of studies of hematologic effects summarized in *Volume 1* are inconsistent.

APPLICATION OF THE TOXICOLOGIC DATA

As discussed in this chapter, animal toxicity studies have been conducted primarily in rats, mice, and dogs and to a smaller extent in monkeys. The studies exposed groups of laboratory-bred animals to different concentrations of uranium compounds for various portions of the animals' life span by different routes (for example, inhalation, ingestion via drinking water, and surgical implantation). Such experiments provide information that is necessary to determine short-term or longer-term effects on the body, specific organ systems, and biochemical processes. The advantage of the studies is that the conditions of dose and exposure are carefully monitored to maintain control over many of the experimental characteristics and to minimize confounders and so allow scrutiny of the specific effects of the uranium or depleted uranium. The disadvantage is that the animals are not the primary species of interest for human toxicity. Toxicity in animals is not always predictive of effects in humans.

Toxicologic studies (animal and in vitro studies) typically involve administration of high doses of a test substance, in this case uranium, that are generally greater than those received by humans. For example, in the animal cancer study by Mitchel et al. (1999) discussed above, average urinary uranium concentrations were 0.117-0.274 mg/L compared with a study of uranium mill workers in which the mean urinary uranium concentrations were 0.0652 mg/L in 1975 and 0.0072 mg/L in 1981 (Thun et al., 1985) and a study of uranium in drinking water in which the mean concentration in residents was 0.000424 mg/L (Kurttio et al., 2002). In addition, the urinary uranium concentration in the 50th percentile of the US population is 6.32 ng/L (Ting et al., 1999). The large difference makes it difficult to extrapolate from effects observed in animals to health outcomes in humans. Toxicologic studies conducted using lower doses (similar to doses received by humans) may provide more relevant information about the relationship between effects in animals and human health outcomes.

For the reasons discussed above, the committee considered toxicologic studies to be secondary information sources. Information from such studies was used to determine mechanism of action but not to determine human health outcomes.

TABLE 3-1 Uranium Compounds, by Dissolution Type

Type F (Fast)	Type M (Medium)	Type S (Slow)
Uranium hexafluoride (UF_6)	Uranium tetrafluoride (UF_4)	Uranium dioxide (UO_2)
Uranium tetrachloride (UCl_4)	Uranium trioxide (UO_3)	Triuranium octaoxide (U_3O_8)
Uranyl fluoride (UO_2F_2)	Uranyl acetate ($UO_2(CH_3CO_2)_2$	Uranium peroxide (UO_4)
Uranyl nitrate hexahydrate [$UO_2(NO_3)_2{\cdot}6H_20$]		

TABLE 3-2 Carcinogenic Effects

Species	Route of Exposure and Dose	Frequency and/or Duration	Outcome(s)	Reference
Rat	90% enriched ^{235}U as tetravalent (uranium, 0.57-18.7 mg U/kg of body weight) or hexavalent (uranium, 0.55-5.32 mg U/kg of body weight) U Intratracheal injection	Not known	Statistically significantly increased incidence of osteosarcoma, lung and kidney carcinoma, lung reticulolymphosarcoma, leukemia in treated animals compared with controls	Filippova et al., 1978
Rat, male Sprague Dawley	Natural uranium-ore dust aerosol, 50 mg/m^3 or 19 mg/m^3 Nose-only inhalation	4.2 hours/day, 5 days/week for 65 weeks	Frequency of primary malignant lung tumors 0.016, 0.175, and 0.328 and frequency of primary nonmalignant lung tumors 0.016, 0.135, and 0.131 in control, low-exposure, and high-exposure groups, respectively	Mitchel et al., 1999
Rat, male Wistar	DU pellets: 2.0 mm × 1.0 mm in diameter, 6.0 Bq[a] DU fragments: 2.5 × 2.5 × 1.5 mm, 20 Bq[a]; 5.0 × 5.0 × 1.5 mm, 59 Bq[a] 4 implants/rat, intramuscular	Observed for lifetime	Significant increase in incidence of soft-tissue sarcoma in 5.0 × 5.0 × 1.5-mm group; slight increase in 2.5 × 2.5 × 1.5-mm group; no increase in 2.0 × 1.0-mm group	Hahn et al., 2002

Continued

TABLE 3-2 Continued

Species	Route of Exposure and Dose	Frequency and/or Duration	Outcome(s)	Reference
Mouse, male DBA/2	DU pellets: size and radioactivity not specified 2, 6, 8 implants/mouse, intramuscular 10^6 FDC-P1 hematopietic cells administered 60 days after pellet implantation		76% of treated mice developed leukemias compared with 12% of controls	Miller et al., 2005
Hamster, golden Syrian	Uranium-ore dust, 19 mg/m^3	16 months	No increase in number of tumors in treated animals compared with controls	Cross et al., 1981
Monkey and dog	Uranium dioxide aerosol, 5 mg/m^3 Inhalation	Up to 5 years	Frank neoplasms and foci of atypical epithelial proliferation in 31% and 46%, respectively, of surviving dogs kept 75 months after termination of 5-year exposure; pulmonary tumors and atypical epithelial changes not found in any exposed monkeys	Leach et al., 1973

[a]Effective alpha-particle radioactivity emanating from the surface of the DU.
NOTE: DU = depleted uranium.

TABLE 3-3 Genotoxic Effects

System Studied	Route of Exposure/Dose	Frequency and/or Duration	Outcome(s)	Reference
Human				
34 Gulf War veterans (BVAMC DU Followup Program)	High-DU group (n = 10), >0.1 μg/g creatinine; low-DU group (n = 24), <0.1 μg/g creatinine	12- and 14-year postwar followup	CA assay: no significant differences between groups HPRT mutation frequency: no significant differences between groups SCE assay: no association between SCE and urinary DU concentration (assayed at 12-year followup only)	McDiarmid, 2007
39 Gulf War veterans (BVAMC DU Followup Program)	Urinalysis results: 0.001-78.125 μg/g creatinine High-DU group (n = 13), >0.1 μg/g creatinine; low-DU group (n = 26), <0.1 μg/g creatinine	10-year postwar followup	CA assay: high-DU group had higher CA frequency per cell SCE assay: no association between SCE and urinary DU concentration HPRT mutation frequency: statistically significant positive association between HPRT mutation frequency and urinary DU concentration	McDiarmid et al., 2004
50 Gulf War veterans (BVAMC DU Followup Program)	Urinalysis results: 0.002-31.8 μg/g creatinine High-DU group (n = 13), >0.1 μg/g creatinine; low-DU group (n = 37), <0.1 μg/g creatinine	8-year postwar followup	CA assay: no association between CA and urinary DU concentration SCE assay: statistically significant increase in baseline SCE in high-exposure group compared with low-exposure group (P = 0.03)	McDiarmid et al., 2001

Continued

TABLE 3-3 Continued

System Studied	Route of Exposure/Dose	Frequency and/or Duration	Outcome(s)	Reference
69 people in south of Serbia and Montenegro (where DU ammunition was used during 1999 air strikes)	Urinalysis results: alpha-spectrometry (range): 1-30.4 mBq/L in exposed group versus "below detection" in controls	1999-2002	CA assay: non-statistically significant increase in CA in exposed group (residents of Vranje and Bujanovac) compared with controls; increase was below incidence of CA in people occupationally exposed to ionizing radiation	Milacic et al., 2004
30 people in Hadžiči (near Sarajevo) environmentally exposed to DU	Urinalysis not conducted Water from two local wells had traces of DU (0.38μg/L [14% of total uranium was DU] and 0.55 μg/L [73.4% of total uranium was DU])	Blood samples taken in 2002, 2003	Micronucleus cytochalasin-B test: statistically significant increase in micronucleus frequencies in exposed group compared with controls	Krunic et al., 2005
Whole animal exposure				
Sprague Dawley, male	Inhalation exposure: UO_4: 116 ± 60 mg/m^3 UO_2 + UO_4: 375 ± 70 mg/m^3 + 116 ± 60 mg/m^3 UO_2 + UO_4: 190 ± 41 mg/m^3 + 116 ± 60 mg/m^3	UO_4: 30 minutes UO_2 + UO_4: 3 hours + 30 minutes UO_2 + UO_4: 3 hours 4 days/week for 3 weeks + 30 minutes	Comet assay: UO_4 exposure alone had no effect on DNA damage; repeated UO_2 pre-exposure followed by UO_4 exposure increased DNA damage compared with controls	Monleau et al., 2006b

Sprague Dawley, male	Nose-only inhalation exposure: UO_2: 190 ± 41 mg/m^3 or 375 ± 70 mg/m^3 UO_4: 116 ± 60 mg/m^3	UO_2: 30 minutes (190 mg/m^3), acute and repeated exposure; 2 hours (375 mg/m^3); or 3 hours (375 mg/m^3) UO_4: 30 minutes	Comet assay: DNA damage occurred in groups exposed to UO_2 at single 375-mg/m^3 dose for 3 hours and repeated 190-mg/m^3 dose; no DNA damage in other groups	Monleau et al., 2006c
Sprague Dawley, male	Pellets 1 mm in diameter × 2 mm long; implanted in gastrocnemius muscle; low-, medium-, and high-dose groups	Urine, serum samples collected 6, 12, 18 months after pellet implantation	Ames *Salmonella* reversion assay: increased urinary uranium content led to increased mutagenicity	Miller et al., 1998b
In vitro studies				
Human bronchial fibroblast cell line that ectopically expresses human telomerase, WTHBF-6	Uranyl acetate: 100, 200, 400, 800 μM Uranium trioxide: 0.5, 1, 5, 10 μg/cm^3	Cells incubated for 24, 48, 72 hours	Both compounds led to time- and concentration-dependent cytotoxicity; uranium trioxide led to increased chromosomal damage, but uranyl acetate did not	Wise et al., 2007
Human liver carcinoma cells (HepG2)	DU-UO_2 at 0-50 μg/mL	Cells incubated for 48 hours	DU exposure led to dose-dependent induction of nine of 13 promoters assayed	Miller et al., 2004
Human osteoblast cells (HOS)	Induction assay: 0-50 μM DU-uranyl nitrate Transformation assay: 50 μM DU-uranyl nitrate (46 cGy alpha-particle equivalent dose), ^{238}U-uranyl nitrate (35 cGy), ^{235}U-uranyl nitrate (227.5 cGy)	Cells incubated for 24 hours	Induction assay: DU exposure led to dose-dependent increase in yield of dicentrics Transformation assay: DU exposure led to specific activity-dependent increase in neoplastic transformation frequency	Miller et al., 2002b

Continued

TABLE 3-3 Continued

System Studied	Route of Exposure/Dose	Frequency and/or Duration	Outcome(s)	Reference
Human osteoblast cells (HOS)	DU-UO_2 at 5 or 10 mg/mL	Cells incubated for 24 hours	Transformation assay: DU exposure at 10 mg/mL led to 25.5-fold increase in transformation frequency compared with untreated HOS cells Genotoxicity assays (micronuclei induction, SCE concentration, DNA single-strand breaks, dicentric formation): DU exposure at 5 mg/mL led to significant increases compared with untreated cells	Miller et al., 2002c
Human osteoblast cells (HOS)	DU-UO_2Cl_2 at 10 μg/mL DU-UO_2 or 10 μM	Cells incubated for 24 hours	DU exposure led to increase in transformation to tumorigenic phenotype; cells not transformed after exposure to DU and phenyl acetate (RAS protein target)	Miller et al., 2001
Human osteoblast cells (HOS)	DU-UO_2Cl_2 at 10 μM	Cells incubated for 24 hours	SCE assay: DU exposure led to about 2-fold increase in SCE induction Transformation assay: DU exposure led to 9.6-fold increase in transformation frequency compared with controls	Miller et al., 1998a

Chinese hamster lung fibroblast V79 cells	DU-uranyl nitrate at 10-50 μg/mL	Cells incubated for 24 hours	Dose-dependent increase in mutagenic response after exposure; at equal uranium concentration, higher specific activity led to increase in *hprt* mutant frequency suggesting that radiation is involved in DU-induced biological effects in vitro	Miller et al., 2007
Chinese hamster ovary (CHO) EM9 cells	200 μM UA	Cells incubated for 24 hours	UA-exposed cells had statistically significantly more genomic mutations than controls	Coryell and Stearns, 2006
Chinese hamster ovary (CHO) EM9 cells	0-300 μM UA	Cells incubated for 40 minutes (comet assay), 24 hours (cytotoxicity, mutagenicity, comet assays; measurement of uranium-DNA-P binding), 48 hours (measurement of uranium-DNA-P binding)	Cytotoxicity assay: UA exposure led to more cytotoxicity in EM9 cells (which are DNA-repair-deficient) compared with controls (3.1-fold increase in cell death at 200 μM) Mutagenicity assay: UA exposure led to higher induced mutant frequency (about 5-fold) in EM9 cells compared with controls Comet assay: no differences in tail moments between EM9 cells and controls Uranium/DNA-P binding: No significant difference in uranium-DNA adduct between EM9 and control cells	Stearns et al., 2005

Continued

TABLE 3-3 Continued

System Studied	Route of Exposure/Dose	Frequency and/or Duration	Outcome(s)	Reference
Normal rat renal proximal cells	Up to 700 μM uranium biocarbonate	Cells incubated for 24 hours	DNA damage occurred in time- and concentration-dependent manner; exposure at 300 μM or higher led to genotoxicity; DNA damage may be reversible at low concentrations and irreversible at higher concentrations	Thiébault et al., 2007
pBluescript SK DNA+ plasmid DNA	0.1-1.0 mM UA	30 minutes at 37°C	UA + ascorbic acid exposure led to single-strand breaks, demonstrating chemical genotoxicity	Yazzie et al., 2003
Calf thymus DNA	1-1,000 μM DU-uranyl nitrate	30 minutes at 37°C	DU exposure led to oxidative DNA damage without significant alpha-particle decay	Miller et al., 2002a

NOTE: BVAMC = Baltimore Veterans Affairs Medical Center, CA = chromosomal aberration, DU = depleted uranium, *hprt* = hypoxanthine-guanine phosphoribosyl transferase, SCE = sister chromatid exchange, UA = uranyl acetate, UO_2 = uranium dioxide, UO_2Cl_2 = uranyl chloride, UO_4 = uranium peroxide.

TABLE 3-4 Respiratory Effects

Cell Line	Dose	Duration	Outcomes	Reference
Human type II epithelial cell line	0.5 mM uranyl acetate	48-hour incubation	Exposure triggered differential expression of 18 spots, of which 14 corresponded to fragments of cytokeratin 8 (CK8) and cytokeratin 18 and one to peroxiredoxin 1; CK cleavage did not result from caspase or calpain activity	Malard et al., 2005
Rat lung epithelial cell line	0.25, 0.5, 1 mM uranyl acetate	3-hour incubation	Induction of oxidative stress at 0.5 and 1 mM; response correlated with dose and time	Periyakaruppan et al., 2007
Rat pulmonary alveolar macrophage cell line, NR8383	10-300 μM uranyl acetate	24-hour incubation	Dose-dependent increase in TNFα production; no secretion on IL-1β, IL-10 detected	Gazin et al., 2004

NOTE: IL = interleukin, TNF = tumor necrosis factor.

TABLE 3-5 Renal Effects

Species	Route of Exposure and Dose	Frequency or Duration	Outcomes	Reference
Sprague Dawley rat, male	Ingestion via drinking water Uranium nitrate at 40 mg/L (14.5 Bq)	6, 9, 12, 18 months	Exposed rats had increased number of vesicles containing dense granular inclusions in proximal tubular cells; inclusions composed of small granule clusters and increased in number with exposure duration; granules composed of iron oxides	Donnadieu-Claraz et al., 2007
Wistar rat, male	Intramuscular injection DU nitrate at 0.2, 1.0, 2.0 mg/kg	Animals euthanized 28 days after exposure	DU concentrations in urine decreased within 3-7 days after exposure; NAG-creatinine concentrations peaked at day 3 with high correlation to injected DU doses	Fukuda et al., 2006
BBMV from unspecified rat species	Uranyl acetate at 0.25, 0.5, or 1.0 mg/mg of protein of BBMV	30 min	Uranyl acetate exposure led to decrease in glucose transport	Goldman et al., 2006
New Zealand White rabbit, males and females	Uranyl nitrate hexahydrate at 0.96, 4.8, 24, 120, 600 mg/L (0.49-43.02 mg/kg body weight for females; 0.05-28.7 mg/kg body weight for males)	91 days	Histopathological changes observed in kidneys of animals exposed at all doses	Gilman et al., 1998b
Dogs and monkeys	Inhalation Uranium dioxide at 5.8 mg/m^3	5.4 hours/day, 5 days/week, up to 5 years	Injury to kidneys not observed as result of exposure	Leach et al., 1970

NOTE: BBMV = brush border membrane vesicles, DU = depleted uranium, NAG = N-acetyl-β-D-glucosaminidase.

TABLE 3-6 Neurologic Effects

Species or Cell Line	Route of Exposure and Dose	Frequency or Duration	Outcomes	Reference
Rat				
Sprague Dawley, male	Inhalation: uranium dioxide (1.3×10^4 Bq/g, 190 mg/m^3) Ingestion (drinking water): enriched uranium (6.63×10^4 Bq/g) or DU (1.47×10^4 Bq/g) Intraperitoneal injection: enriched uranium (16.5 Bq) + ^{233}U (50 Bq)	Inhalation: 30 minutes/day, 4 days/week for 3 weeks Ingestion: 1 mg/day for 42 days Injection: single injection	Overall, amount of uranium entering brain was low; location of accumulation in brain depended on route of exposure: in injection group, uranium distributed among different cerebral areas; in ingestion group, triata had highest uranium concentration; in inhalation group, olfactory bulbs had highest uranium concentration; high accumulation found in hippocampus by each route of exposure	Houpert et al., 2007c
Sprague Dawley, male	Ingestion (drinking water) Enriched uranium nitrate at 40 mg/L	3, 6, 9 months	Exposure to enriched uranium for 3, 9 months significantly reduced spontaneous alternation measured in Y-maze; no differences between exposed and control rats in the open-field, object-recognition, forced-swimming tests	Houpert et al., 2007a
Sprague Dawley, primary rat cortical neurons	Cell culture 1, 10, or 100 μM uranyl acetate	Cells treated 3 weeks after isolation for 24 hours	Exposure did not lead to significant changes in cellular energy metabolism, thiol metabolite oxidation, lipid metabolism	Jiang et al., 2007

Continued

TABLE 3-6 Continued

Species or Cell Line	Route of Exposure and Dose	Frequency or Duration	Outcomes	Reference
Sprague Dawley, male	Ingestion (drinking water) Uranyl nitrate at 40 mg/L (daily intake about 4-1.5 mg/kg of body weight)	1.5, 6, 9 months	Exposure to uranyl nitrate for up to 9 months did not affect AChE activity in striatum, hippocampus, frontal cortex; exposure for 6 months altered AChE activity in cerebellum; exposure for 6 months led to small change in DAergic turnover ratio in frontal cortex; after 9 months, significant decrease in 5HIAA concentration and 5HTergic turnover ratio in frontal cortex, decrease in the DOPAC concentration and DAergic turnover ratio in striatum	Bussy et al., 2006
Rat brain endothelial cells (RBE4)	10, 50, 100 μM U_3O_8 uranyl chloride	15, 30 minutes	No overt cytotoxicity observed with assays for cell-volume increase, heat-shock protein 90 expression, 3-[4,5-dimethylthiazol-2-yl]-2,5-diphenyltetrazolium bromide reduction, lactate dehydrogenase activity	Dobson et al., 2006
Sprague Dawley, male	Implanted in gastrocnemius muscle DU pellets: 1 mm in diameter × 2 mm long	3, 6 months	At 3 months, DU accumulated in all brain regions except the hippocampus of highest-dose animals; at 6 months, DU accumulation found in cortex, midbrain, cerebellum	Fitsanakis et al., 2006
Sprague Dawley, male	Intraperitoneal injection Uranyl acetate dehydrate at 1mg/kg	Uranyl acetate administered after forced swimming to induce stress; animals euthanized 8 hours, 24 hours, 7 days, 30 days after exposure	Stress increased clearance of uranium in the brain compared with clearance of brain uranium in unstressed rats	Barber et al., 2005

Long Evans, male and female	Ingestion (drinking water) DU acetate dehydrate at 75, 150 mg/L (25, 50 mg/kg per day)	2 weeks, 6 months	Exposure for 2 weeks led to behavioral changes (line-crossing, rearing) in males and changes in brain lipids in both sexes compared with controls; exposure for 6 months produced additional behavioral changes	Briner and Murray, 2005
Sprague Dawley, sex not specified	Ingestion (drinking water) Enriched or depleted uranium incorporated as nitrate: 1 mg/day per rat	1.5 months	Exposure to enriched uranium led to increases in amount of paradoxical sleep and anxiety and reduction in spatial working-memory capacities compared with controls; exposure to DU did not lead to these effects	Houpert et al., 2005
Sprague Dawley, male	Intraperitoneal injection Uranyl nitrate at 70, 144 μg/kg	Animals euthanized 3 days after exposure	Exposure at higher concentration led to shorter paradoxical sleep compared with controls; no effect at lower concentration	Lestaevel et al., 2005b
Sprague Dawley, male	Ingestion (drinking water) Uranyl nitrate at 40 mg/L	3 months	By 30 days, exposure to uranium led to increase in rapid-eye-movement sleep and theta-band power during light period	Lestaevel, 2005a
Sprague Dawley, male	Nose-only inhalation Uranium dioxide at 197 mg/m^3	30 minutes/day, 4 days/week for 3 weeks	Exposure led to behavioral changes (spontaneous locomotion activity, spatial working memory) compared with controls	Monleau et al., 2005
Sprague Dawley, male	Ingestion (drinking water) DU at 40 mg/L	30 days	Exposure to DU did not lead to alteration of dopamine, serotonin, catabolite concentrations in striatum, hippocampus, cerebral cortex, thalamus, cerebellum	Houpert et al., 2004

Continued

TABLE 3-6 Continued

Species or Cell Line	Route of Exposure and Dose	Frequency or Duration	Outcomes	Reference
Sprague Dawley, male	In situ brain perfusion 5×10^{-8} to 5×10^{-5} M uranyl tricarbonate	2 minutes	Significant amount of uranium found in brains of exposed rats	Lemercier et al., 2003
Sprague Dawley, male	Intramuscular injection Uranyl acetate at 0.1, 1 mg/kg	1 injection per day for 7 days, followed by 30-day observation period	Dose-related deficit in inclined-plane performance; reduced grip time, impaired beam-walk score and beam-walk time at both doses; significant increase in nitric oxide at 0.1 mg/kg in cortex, midbrain; significant increase at 1 mg/kg in AChE activity in the cortex; no change in ligand-binding densities for m2 muscarinic receptor	Abou-Donia et al., 2002
Sprague Dawley, male	Implanted in gastrocnemius muscle DU pellets: 1 mm in diameter × 2 mm long, up to 10 pellets per thigh	Animals euthanized 6, 12, 18 months after implantation	Exposure to DU led to neurophysiologic changes in hippocampus (EPSP-spike coupling); changes occurred at 6, 12 months, but not after 18 months	Pellmar et al., 1999b
Sprague Dawley, sex not specified	Implanted in gastrocnemius muscle DU pellets: 1 mm in diameter × 2 mm long, up to 10 pellets per thigh	Animals euthanized 1, 6, 12, 18 months after implantation	Significant amounts of DU found in motor cortex, frontal cortex, midbrain, cerebellum, vermis compared with controls	Pellmar et al., 1999a
Sprague Dawley, male and female	Ingestion (drinking water) Uranyl nitrate hexahydrate at up to 600 mg/L (37, 54 mg U/kg body weight per day for male, female rats, respectively)	Up to 91 days	No signs of neurotoxicity related to exposure	Gilman et al., 1998c

Sprague Dawley, male	Intragastric Uranyl acetate dehydrate at 11-717 mg U/kg body weight	Single dose	Acute cholinergic toxicity observed in treated rats	Domingo et al., 1987
Mouse				
Phrenic nerve diaphragm from ICR strain, male and female	0.2-0.8 mM uranyl nitrate	More than 4-hour incubation	Treatment facilitated release of acetylcholine from nerve terminals, potentiated muscle contraction	Lin et al., 1988
Other species				
Caenorhabditis elegans (nematode)	1, 10, 100 mM uranyl acetate	24-hour incubation period	Exposure did not lead to significant neurodegeneration	Jiang et al., 2007
Dog	Inhalation Uranium hexafluoride gas at 0.5-18 mg U/m^3	30 days	Muscle weakness and instability of gait beginning on day 13 at 18 mg U/m^3	Dygert et al., 1949
Dog	Inhalation Uranyl nitrate hexahydrate at 9.5 mg U/m^3	30 days	Exposure associated with anorexia	Roberts, 1949
Cat	Inhalation Uranium hexafluoride gas at 0.5-18 mg U/m^3	30 days	Muscle weakness and instability of gait beginning on day 7 at 18 mg U/m^3	Dygert et al., 1949

NOTE: 5HIAA = 5-hydroxyindoleacetic acid, 5HTergic = serotoninergic, AChE = cholinergic acetylcholinesterase, DAergic = dopaminergic, DOPAC = 3,4 dihydroxyphenylacetic acid, DU = depleted uranium.

TABLE 3-7 Hepatic Effects

Species	Route of Exposure and Dose	Frequency or Duration	Outcomes	Reference
Rat				
Sprague Dawley rat, male	Subcutaneous administration Uranyl hexahydrate at 11.5 mg/kg	Single administration Animals euthanized 1, 3 days after exposure	Exposure led to changes in hepatic metabolism of bile acids and xenobiotics; DU appeared to act through modulation of CYP enzymes	Guegeun et al., 2006
Sprague Dawley rat, male	Ingestion (drinking water) Uranyl nitrate at 1 mg/day per animal	9 months, followed by intraperitoneal injection of acetaminophen at 400 mg/kg 2, 24 hours before euthanasia	Plasma acetaminophen concentrations higher in DU group treated for 24 hours compared with non-DU group; DU-treated group had significantly increased ALT, AST after 2 hours of acetaminophen treatment and decreased CYP2A, 2B, 3A activity compared with controls	Gueguen et al., 2007
Sprague Dawley rat, male	Ingestion (drinking water) Uranyl nitrate: 1 mg/day per animal	9 months	mRNA concentrations of CYP3A1, CYP3A2, PXR significantly higher in the DU-exposed group compared with controls; mRNA concentrations of CYP1A1, CYP2B1, RXR, CAR unchanged; hepatic activity of CYP2A, CYP2B, CYP2C, CYP3A did not change significantly in treated vs control groups	Souidi et al., 2005
Other species				
Danio rerio (zebrafish), male adults	DU: 1.5 Bq/L DU + ^{233}U: 2376 Bq/L	3, 10, 20 days	Decreased superoxide dismutase, catalase activities, total glutathione content in liver extracts	Barillet et al., 2007

NOTE: ALT = alanine amino transferase, AST = aspartate amino transferase, CYP = cytochrome P450, DU = depleted uranium, i.p. = intraperitoneal, mRNA = messenger RNA.

TABLE 3-8 Reproductive and Developmental Effects

Species	Route of Exposure and Dose	Frequency or Duration	Outcomes	Reference
Rat				
Sprague Dawley, male and female	Ingestion in drinking water Enriched uranium nitrate at 40 mg/L (about 1 mg/day per rat) (P1 generation)	Exposure began 3 months before mating and continued through lactation	Exposure to uranium led to delayed hyperactivity, behavioral decrements in performance (open-field, Y-maze, elevated plus-maze tests) in 2-, 5-, 9-month-old offspring	Houpert et al., 2007b
Sprague Dawley, male and female	Implanted in gastrocnemius muscle DU pellets: 1 mm in diameter × 2 mm long, up to six pellets per calf (P1 generation)	P1 generation mated 30 days after implantation and monitored during pregnancy; offspring (F1 generation) assessed through adult stage	Exposure to DU did not lead to reproductive effects in P1 generation compared with controls; no developmental effects (birth weight, survival, litter size, gross physical abnormalities, neurodevelopmental effects, immune-function effects) in F1 generation	Arfsten et al., 2005
Sprague Dawley, male and female	Implanted in gastrocnemius muscle DU pellets: 1 mm in diameter × 2 mm long, up to 10 pellets per calf (P1 generation)	P1 generation mated 30, 120 days after implantation	Exposure to DU did not adversely affect male reproductive success, sperm concentration, sperm velocity compared with controls	Arfsten et al., 2006
Sprague Dawley, male and female	Subcutaneous injection Uranyl acetate dehydrate at 0.415, 0.830 mg/kg per day	Administered on gestation days 6-15 with or without restraint stress for 2 hours/day; cesarean sections performed on gestation day 20	Exposure to DU led to maternal toxicity and embryotoxicity in high-dose group; fetotoxicity (reduction in fetal body weight, increase in total number of skeletally affected fetuses) observed in both groups; no teratogenic effects observed in either group; maternal restraint stress enhanced embryo, fetal toxicity in high-dose group	Albina et al., 2003

Continued

TABLE 3-8 Continued

Species	Route of Exposure and Dose	Frequency or Duration	Outcomes	Reference
Mouse				
$B6C3F_1$, male and female C57Bl/6, ovariectomized females	Ingestion in drinking water Uranium hexahydrate at 0.5 μg/L (0.001 μM) to 28 mg/L (120 μM)	Experiment 1: administered for 30 days to immature 28-day-old mice ($B6C3F_1$) Experiment 2: administered to males and females for 30 days before breeding, then to pregnant females through gestation ($B6C3F_1$) Experiment 3: administered for 30 days starting 7 days after ovarectomy (C57Bl/6) Experiment 4: administered uranium or DES for 10 days beginning at age of 50 days (C57Bl/6), intraperitoneal injection of ICI 182,780	Exposure to uranium hexahydrate led to estrogenic responses (selective reduction in primary follicles, increase in uterine weight, greater uterine luminal epithelial cell height, accelerated vaginal opening, persistent presence of cornified vaginal cells); pups exposed in utero had significantly fewer primordial follicles than unexposed	Raymond-Whish et al., 2007
Swiss Webster, female	Ingestion in drinking water Uranium acetate at 19, 37, 75 mg/L	Exposed for 2 weeks, then mated; exposure of dams and offspring continued until sacrifice	Exposure to uranium acetate did not lead to maternal toxicity; no gross malformations observed in pups Exposed offspring developed more quickly than controls on behavior indexes (righting reflexes, forelimb placing and grasping, swimming development); hindlimb placing	Briner and Byrd, 2000

			was at first accelerated in exposed group, but timepoint was delayed later; on functional observation battery, exposed offspring had fewer spontaneous vocalizations and on touch-response test, more freezing and jerking behavior; exposed offspring gained weight more quickly than controls and at sacrifice had significantly higher body and brain weights, although brain as percentage of body weight was smaller in exposed groups (37 and 75 mg/L)	
Swiss, male	Ingestion (drinking water) Uranyl acetate dihydrate: 0, 10, 20, 40, or 80 mg/kg/day	64 days	Testicular function and spermatogenesis were not affected by exposure to uranium	Llobet et al., 1991
Swiss, male and female	Intragastic administration Uranyl acetate dihydrate: 0, 5, 10, 25 mg/kg/day	Males: 60 days prior to mating Females: 60 days prior to mating and throughout mating, gestation, parturition, and nursing	Uranium exposure did not lead to adverse effects on fertility; increased embryolethality was observed in the highest dose group	Paternain et al., 1989

NOTE: DES = diethylstilbestrol, DU = depleted uranium, ICI 182,780 = a steroidal estrogen antagonist.

TABLE 3-9 Immunologic Effects

Species or Cell Line	Route of Exposure and Dose	Frequency or Duration	Outcomes	Reference
Animal studies				
Sprague Dawley, male	Nose-only inhalation UO_2 at 190 ± 41 mg/m^3, 375 ± 70 mg/m^3 UO_4 at 116 ± 60 mg/m^3	UO_2: 30 minutes (190 mg/m^3), acute and repeated exposure; 2 hours (375 mg/m^3); or 3 hours (375 mg/m^3) UO_4: 30 minutes	Inflammatory cytokine expression increased	Monleau et al., 2006c
Sprague Dawley, male	Ingestion in drinking water Uranyl nitrate at 40 mg/L (1 mg/day per animal)	3, 6, 9 months	DU exposure led to decrease in intestinal mast cell number, increase in IL-1β and IL-10 concentrations, decrease in mRNA CCL-2 concentrations, decrease in intestinal macrophage density, increase in number of neutrophils	Dublineau et al., 2007
Sprague Dawley, male	Gavage Uranyl nitrate at 204 mg/kg	Single administration Animals euthanized 1 or 3 days after administration	DU appeared to modulate the expression and/or production of IFNγ, MCP-1 in intestine	Dublineau et al., 2006b
Sprague Dawley, male (harvested intestines)	Ingestion in drinking water Uranyl nitrate at 1 mg/day per animal	3, 9 months	DU preferentially accumulated in Peyer's patches compared with epithelium; no induction of apoptosis pathway after chronic DU contamination in Peyer's patches; no change in cytokine expression (Il-10, TGF-β, IFN-γ, TNF-α, MCP-1) in Peyer's patches and mesenteric lymph nodes; no modification in uptake of yeast cells by Peyer's patches	Dublineau et al., 2006a

Sprague Dawley, male and female	DU pellets: 1 × 2 mm 0, 4, 8, 12 implants/rat (P1 generation)	P1 generation mated 30 days after surgery; P1 females nursed offspring until PND 20	No significant difference among 8-week-old F1 treatment groups in mean thymus and spleen mass, mean total number of thymocytes per spleen	Arfsten et al., 2005
Sprague Dawley, sex not specified	Implanted in gastrocnemius muscle DU pellets: 1 mm in diameter × 2 mm long; up to 20 pellets per thigh	1 day, 1, 6, 12, 18 months	Significant concentrations of uranium found in spleen	Pellmar et al., 1999a
In vitro studies				
Macrophages from BALB/c and DO11.10 T-cell receptor mice	Uranyl nitrate: 10-1,000 μM	Cells incubated for 2 hours	Lymphoproliferation assay: uranyl nitrate exposure at 200 μM for 2 hours led to altered macrophage accessory-cell function	Wan et al., 2006
Macrophage cell line, J774	DU-uranyl chloride: 1, 10, 100 μM	Cells incubated for up to 24 hours	Exposure at all concentrations led to decreased viability of cells; appeared to be apoptotic death	Kalinich et al., 2002

NOTE: CCL = chemokine ligand 2, DU = depleted uranium, IFN = interferon, IL = interleukin, MCP = monocyte chemoattractant protein, PND = post natal day, TGF = transforming growth factor, TNF = tumor necrosis factor, UO_2 = uranium dioxide, UO_4 = uranium peroxide.

TABLE 3-10 Musculoskeletal Effects

Species	Route of Exposure and Dose	Frequency or Duration	Outcomes	Reference
Rat				
Sprague Dawley, male	Ingestion in drinking water Uranyl nitrate at 1 mg/day per animal	9 months	Decrease in vitamin D concentration in plasma compared with controls; expression of CYP genes involved in vitamin D metabolism unaltered in liver of treated animals; significant decrease in *cyp24a1* mRNA concentrations in kidneys of treated animals	Tissandie et al., 2007
Sprague Dawley, male	Gavage Uranyl nitrate at 204 mg/kg of body weight (LD_{50} at 14 days)	Single dose Animals were euthanized 1 or 3 days after exposure	Significant decreases in vitamin D and PHT in plasma compared with controls; treatment modulated mRNA concentrations and activity of CYP enzymes involved in vitamin D metabolism	Tissandie et al., 2006
Wistar, sex not specified	Oral Uranyl nitrate at 90 mg/kg of body weight	Single dose Animals were euthanized 7 or 27 days after exposure	No statistically significant difference in mandibular length observed in treated vs control animals; mandibular area and height, tooth eruption, dental development decreased in treated animals at 7 days but similar to controls by 27 days	Pujadas Bigi and Ubios, 2007
Wistar, male	Intramuscular injection DU nitrate at 0.2, 1.0, 2.0 mg/kg of body weight	Single dose Animals euthanized 28 days after exposure	Increases in concentrations of osteocalcin, tartrate-resistant acid phosphatase, pyridinoline, parathyroid hormone in all treated groups compared with controls	Fukuda et al., 2006

Sprague Dawley, male and female	Ingestion in drinking water Uranyl nitrate hexahydrate at 0.96, 4.8, 24, 120, 600 mg/L (0.09-53.56 mg/kg of body weight for females; 0.06-36.73 mg/kg of body weight for males)	91 days	No significant exposure-related effect on hematologic, biochemical endpoints	Gilman et al., 1998c
Wistar, male	Intraperitoneal injection Uranyl nitrate at 2 mg/kg	Single dose	Alveolar bone volume (15×10^5 μm^2 vs 34×10^5 μm^2), total bone formation areas (4.85% vs 19.55%), volume density of bone in the alveolar apical third (0.26 vs 0.40) significantly lower in intoxicated animals compared with controls	Guglielmotti et al., 1985
Rabbit				
New Zealand White, males and females	Uranyl nitrate hexahydrate at 0.96, 4.8, 24, 120, 600 mg/L (0.49-43.02 mg/kg of body weight for females; 0.05-28.7 mg/kg of body weight for males)	91 days	No significant exposure-related effect on hematologic, biochemical endpoints	Gilman et al., 1998b

NOTE: DU = depleted uranium, LD_{50} = dose required to kill half the test population, PHT = parathyroid hormone.

REFERENCES

Abou-Donia, M. B., A. M. Dechkovskaia, L. B. Goldstein, D. U. Shah, S. L. Bullman, and W. A. Khan. 2002. Uranyl acetate-induced sensorimotor deficit and increased nitric oxide generation in the central nervous system in rats. *Pharmacology, Biochemistry, and Behavior* 72(4):881-890.

Albina, M. L., M. Belles, M. Gomez, D. J. Sanchez, and J. L. Domingo. 2003. Influence of maternal stress on uranium-induced developmental toxicity in rats. *Experimental Biology and Medicine* 228(9):1072-1077.

Arfsten, D. P., M. Bekkedal, E. R. Wilfong, J. Rossi 3rd, K. A. Grasman, L. B. Healey, J. M. Rutkiewicz, E. W. Johnson, A. R. Thitoff, A. E. Jung, S. R. Lohrke, D. J. Schaeffer, and K. R. Still. 2005. Study of the reproductive effects in rats surgically implanted with depleted uranium for up to 90 days. *Journal of Toxicology and Environmental Health A* 68(11-12):967-997.

Arfsten, D. P., D. J. Schaeffer, E. W. Johnson, J. Robert Cunningham, K. R. Still, and E. R. Wilfong. 2006. Evaluation of the effect of implanted depleted uranium on male reproductive success, sperm concentration, and sperm velocity. *Environmental Research* 100(2):205-215.

ATSDR (Agency for Toxic Substances and Disease Registry). 1999. *Toxicological profile for uranium*. Atlanta, GA: US Department of Health and Human Services.

Barber, D. S., M. F. Ehrich, and B. S. Jortner. 2005. The effect of stress on the temporal and regional distribution of uranium in rat brain after acute uranyl acetate exposure. *Journal of Toxicology and Environmental Health A* 68(2):99-111.

Barillet, S., C. Adam, O. Palluel, and A. Devaux. 2007. Bioaccumulation, oxidative stress, and neurotoxicity in Danio Rerio exposed to different isotopic compositions of uranium. *Environmental Toxicology and Chemistry* 26(3):497-505.

Bassett, S. H., A. Frenkel, and N. Cedars. 1948. The excretion of hexavalent uranium following intravenous administration. II. Studies on human subjects. USAEC Report UR-37. As cited in: ATSDR. 1999. *Toxicological Profile for Uranium*. Atlanta, GA: ATSDR.

Berlin, M., and B. Rudell. 1986. Uranium. In *Handbook on the toxicology of metals, 2nd ed.* edited by L. Friberg, G. Nordberg, and V. Vouk. Amsterdam, The Netherlands: Elsevier. Pp. 623-637.

Brady, H. R., B. C. Kone, R. M. Brenner, and S. R. Gullans. 1989. Early effects of uranyl nitrate on respiration and K+ transport in rabbit proximal tubule. *Kidney International* 36(1):27-34.

Brayden, D. J., and A. W. Baird. 1994. A distinctive electrophysiological signature from the Peyer's patches of rabbit intestine. *British Journal of Pharmacology* 113(2):593-599.

Briner, W. 2007. Neurotoxicology of depleted uranium in adult and developing rodents. In *Depleted uranium: Properties, uses, and health consequences*. Boca Raton, FL: CRC Press. Pp. 67-73.

Briner, W., and K. Byrd. 2000. Effects of depleted uranium on development of the mouse. *Metal Ions in Biology and Medicine* 6:459-461.

Briner, W., and J. Murray. 2005. Effects of short-term and long-term depleted uranium exposure on open-field behavior and brain lipid oxidation in rats. *Neurotoxicology and Teratology* 27(1):135-144.

Bussy, C., P. Lestaevel, B. Dhieux, C. Amourette, F. Paquet, P. Gourmelon, and P. Houpert. 2006. Chronic ingestion of uranyl nitrate perturbs acetylcholinesterase activity and monoamine metabolism in male rat brain. *Neurotoxicology* 27(2):245-252.

Cohrssen, J. J., and V. T. Covello. 1989. *Risk analysis: A guide to principles and methods for analyzing health and environmental risks*. Washington, DC: Council on Environmental Quality, Executive Office of the President.

Cooper, J. R., G. N. Stradling, H. Smith, and S. E. Ham. 1982. The behavior of uranium-233 oxide and uranyl-233 nitrate in rats. *International Journal of Radiation Biology and Related Studies in Physics, Chemistry, and Medicine* 41(4):421-433.

Coryell, V. H., and D. M. Stearns. 2006. Molecular analysis of hprt mutations generated in Chinese hamster ovary EM9 cells by uranyl acetate, by hydrogen peroxide, and spontaneously. *Molecular Carcinogenesis* 45(1):60-72.

Cross, F. T., R. F. Palmer, R. H. Busch, R. E. Filipy, and B. O. Stuart. 1981. Development of lesions in Syrian golden hamsters following exposure to radon daughters and uranium ore dust. *Health Physics* 41(1):135-153.

Davies, C. N. 1961. *Inhaled particles and vapours*. Vol 1. London, UK: Pergamon Press.

de Rey, B. M., H. E. Lanfranchi, and R. L. Cabrini. 1983. Percutaneous absorption of uranium compounds. *Environmental Research* 30(2):480-491.

Dobson, A. W., A. K. Lack, K. M. Erikson, and M. Aschner. 2006. Depleted uranium is not toxic to rat brain endothelial (RBE4) cells. *Biological Trace Element Research* 110(1):61-72.

Domingo, J. L., J. M. Llobet, J. M. Tomas, and J. Corbella. 1987. Acute toxicity of uranium in rats and mice. *Bulletin of Environmental Contamination and Toxicology* 39(1):168-174.

Donnadieu-Claraz, M., M. Bonnehorgne, B. Dhieux, C. Maubert, M. Cheynet, F. Paquet, and P. Gourmelon. 2007. Chronic exposure to uranium leads to iron accumulation in rat kidney cells. *Radiation Research* 167(4):454-464.

Dounce, A. L. 1949. The mechanism of action of uranium compounds in the animal body. In *Pharmacology and toxicology of uranium compounds*. Vol. 1, edited by C. Voegtlin and H. C. Hodge. New York: McGraw-Hill. Pp. 951-991.

Dringen, R., and J. Hirrlinger. 2003. Glutathione pathways in the brain. *Biological Chemistry* 384(4):505-516.

Dublineau, I., S. Grison, C. Baudelin, N. Dudoignon, M. Souidi, C. Marquette, F. Paquet, J. Aigueperse, and P. Gourmelon. 2005. Absorption of uranium through the entire gastrointestinal tract of the rat. *International Journal of Radiation Biology* 81(6):473-482.

Dublineau, I., S. Grison, L. Grandcolas, C. Baudelin, C. Tessier, D. Suhard, S. Frelon, C. Cossonnet, M. Claraz, J. Ritt, P. Paquet, P. Voisin, and P. Gourmelon. 2006a. Absorption, accumulation and biological effects of depleted uranium in Peyer's patches of rats. *Toxicology* 227(3):227-239.

Dublineau, I., S. Grison, C. Linard, C. Baudelin, N. Dudoignon, M. Souidi, C. Marquette, F. Paquet, J. Aigueperse, and P. Gourmelon. 2006b. Short-term effects of depleted uranium on immune status in rat intestine. *Journal of Toxicology and Environmental Health A* 69(17):1613-1628.

Dublineau, I., L. Grandcolas, S. Grison, C. Baudelin, F. Paquet, P. Voisin, J. Aigueperse, and P. Gourmelon. 2007. Modifications of inflammatory pathways in rat intestine following chronic ingestion of depleted uranium. *Toxicological Sciences* 98(2):458-468.

Dygert, H. P., C. W. LaBelle, S. Laskin, U. C. Pozzani, E. Roberts, J. J. Rothermel, A. Rothstein, C. J. Spiegl, G. F. J. Sprague, and H. E. Stokinger. 1949. Toxicity following inhalation. In *Pharmacology and toxicology of uranium compounds*, edited by C. Voegtlin and H. C. Hodge. New York: McGraw-Hill. Pp. 423-700.

EPA (Environmental Protection Agency). 2006. *Basic information about radionuclides in drinking water*. http://www.epa.gov/OGWDW/radionuclides/basicinformation.html (accessed March 4, 2008).

Filippova, L. G., A. P. Nifatov, and E. R. Liubchanskii. 1978. [Some of the long-term sequelae of giving rats enriched uranium]. *Radiobiologiia* 18(3):400-405.

Fitsanakis, V. A., K. M. Erikson, S. J. Garcia, L. Evje, T. Syversen, and M. Aschner. 2006. Brain accumulation of depleted uranium in rats following 3- or 6-month treatment with implanted depleted uranium pellets. *Biological Trace Element Research* 111(1-3):185-197.

Frelon, S., P. Houpert, D. Lepetit, and F. Paquet. 2005. The chemical speciation of uranium in water does not influence its absorption from the gastrointestinal tract of rats. *Chemical Research in Toxicology* 18(7):1150-1154.

Fukuda, S., M. Ikeda, M. Chiba, and K. Kaneko. 2006. Clinical diagnostic indicators of renal and bone damage in rats intramuscularly injected with depleted uranium. *Radiation Protection Dosimetry* 118(3):307-314.

Gazin, V., S. Kerdine, G. Grillon, M. Pallardy, and H. Raoul. 2004. Uranium induces TNF alpha secretion and MAPK activation in a rat alveolar macrophage cell line. *Toxicology and Applied Pharmacology* 194(1):49-59.

Gilman, A. P., M. A. Moss, D. C. Villeneuve, V. E. Secours, A. P. Yagminas, B. L. Tracy, J. M. Quinn, G. Long, and V. E. Valli. 1998a. Uranyl nitrate: 91-day exposure and recovery studies in the male New Zealand white rabbit. *Toxicological Sciences* 41(1):138-151.

Gilman, A. P., D. C. Villeneuve, V. E. Secours, A. P. Yagminas, B. L. Tracy, J. M. Quinn, V. E. Valli, and M. A. Moss. 1998b. Uranyl nitrate: 91-day toxicity studies in the New Zealand white rabbit. *Toxicological Sciences* 41(1):129-137.

Gilman, A. P., D. C. Villeneuve, V. E. Secours, A. P. Yagminas, B. L. Tracy, J. M. Quinn, V. E. Valli, R. J. Willes, and M. A. Moss. 1998c. Uranyl nitrate: 28-day and 91-day toxicity studies in the Sprague-Dawley rat. *Toxicological Sciences* 41(1):117-128.

Goldman, M., A. Yaari, Z. Doshnitzki, R. Cohen-Luria, and A. Moran. 2006. Nephrotoxicity of uranyl acetate: Effect on rat kidney brush border membrane vesicles. *Archives of Toxicology* 80(7):387-393.

Gordon, T., and M. O. Amdur. 1991. Responses of the respiratory system to toxic agents. In *Toxicology: The basic sciences of poisons*, edited by M. O. Amdur, J. Doull, and C. D. Klaassen. New York: Pergamon Press. Pp. 383-406.

Gueguen, Y., M. Souidi, C. Baudelin, N. Dudoignon, S. Grison, I. Dublineau, C. Marquette, P. Voisin, P. Gourmelon, and J. Aigueperse. 2006. Short-term hepatic effects of depleted uranium on xenobiotic and bile acid metabolizing cytochrome p450 enzymes in the rat. *Archives of Toxicology* 80(4):187-195.

Gueguen, Y., L. Grandcolas, C. Baudelin, S. Grison, E. Tissandie, J. R. Jourdain, F. Paquet, P. Voisin, J. Aigueperse, P. Gourmelon, and M. Souidi. 2007. Effect of acetaminophen administration to rats chronically exposed to depleted uranium. *Toxicology* 229:62-72.

Guglielmotti, M. B., A. M. Ubios, B. M. de Rey, and R. L. Cabrini. 1984. Effects of acute intoxication with uranyl nitrate on bone formation. *Experientia* 40(5):474-476.

Guglielmotti, M. B., A. M. Ubios, and R. L. Cabrini. 1985. Alveolar wound healing alteration under uranyl nitrate intoxication. *Journal of Oral Pathology* 14(7):565-572.

Guglielmotti, M. B., C. A. Bencini, and A. Keszler. 1987. [Bilateral hemorrhagic bone cyst]. *Revista de la Asociación Odontológica Argentina* 75(5):157-158, 160.

Hahn, F. F., R. A. Guilmette, and M. D. Hoover. 2002. Implanted depleted uranium fragments cause soft tissue sarcomas in the muscles of rats. *Environmental Health Perspectives* 110(1):51-59.

Houpert, P., V. Chazel, F. Paquet, M. H. Henge-Napoli, and E. Ansoborlo. 1999. The effects of the initial lung deposit on uranium biokinetics after administration as UF4 and UO4. *International Journal of Radiation Biology* 75(3):373-377.

Houpert, P., P. Lestaevel, C. Amourette, B. Dhieux, C. Bussy, and F. Paquet. 2004. Effect of U and 137Cs chronic contamination on dopamine and serotonin metabolism in the central nervous system of the rat. *Canadian Journal of Physiology and Pharmacology* 82(2):161-166.

Houpert, P., P. Lestaevel, C. Bussy, F. Paquet, and P. Gourmelon. 2005. Enriched but not depleted uranium affects central nervous system in long-term exposed rat. *Neurotoxicology* 26(6):1015-1020.

Houpert, P., J. C. Bizot, C. Bussy, B. Dhieux, P. Lestaevel, P. Gourmelon, and F. Paquet. 2007a. Comparison of the effects of enriched uranium and 137-cesium on the behaviour of rats after chronic exposure. *International Journal of Radiation Biology* 83(2):99-104.

Houpert, P., S. Frelon, P. Lestaevel, C. Bussy, P. Gourmelon, and F. Paquet. 2007b. Parental exposure to enriched uranium induced delayed hyperactivity in rat offspring. *Neurotoxicology* 28(1):108-113.

Houpert, P., S. Frelon, M. Monleau, C. Bussy, V. Chazel, and F. Paquet. 2007c. Heterogeneous accumulation of uranium in the brain of rats. *Radiation Protection Dosimetry* 127(1-4):86-89.

ICRP (International Commission on Radiological Protection). 1979. *Limits for intakes of radionuclides by workers*. Oxford, UK: Pergamon Press.

———. 2002. *Supporting guidance 3. Guide for the practical application of the ICRP human respiratory tract model*. Oxford, UK: Pergamon Press.

IOM (Institute of Medicine). 2000. *Gulf War and health, volume 1: Depleted uranium, sarin, pyridostigmine bromide, vaccines*. Washington, DC: National Academy Press.

Jiang, G. C., K. Tidwell, B. McLaughlin, J. Cai, R. C. Gupta, D. Milatovic, R. Nass, and M. Aschner. 2007. Neurotoxic potential of depleted uranium—effects in primary cortical neuron cultures and in caenorhabditis elegans. *Toxicological Sciences* 99(2):553-565.

Kalinich, J. F., N. Ramakrishnan, V. Villa, and D. E. McClain. 2002. Depleted uranium-uranyl chloride induces apoptosis in mouse J774 macrophages. *Toxicology* 179(1-2):105-114.

Kathren, R. L., and R. K. Burklin. 2008. Acute chemical toxicity of uranium. *Health Physics* 94(2):170-179.

Krunic, A., S. Haveric, and S. Ibrulj. 2005. Micronuclei frequencies in peripheral blood lymphocytes of individuals exposed to depleted uranium. *Arhiv za Higijenu Rada i Toksikologiju* 56(3):227-232.

Kurttio, P., A. Auvinen, L. Salonen, H. Saha, J. Pekkanen, I. Makelainen, S. B. Vaisanen, I. M. Penttila, and H. Komulainen. 2002. Renal effects of uranium in drinking water. *Environmental Health Perspectives* 110(4):337-342.

Leach, L. J., E. A. Maynard, H. C. Hodge, J. K. Scott, C. L. Yuile, G. E. Sylvester, and H. B. Wilson. 1970. A five-year inhalation study with natural uranium dioxide (UO_2) dust. I. Retention and biologic effect in the monkey, dog and rat. *Health Physics* 18(6):599-612.

Leach, L. J., C. L. Yuile, H. C. Hodge, G. E. S. Iveser, and H.B . Wilson. 1973. A five-year inhalation study with natural uranium dioxide (UO_2) dust. II. Postexposure retention and biologic effects in the monkey, dog and rat. *Health Physics* 25(3):239-258.

Leggett, R. W. 2006. Biokinetics of uranium in the human body. In *Depleted uranium. Properties, uses, and health consequences*, edited by A. C. Miller. London, UK: Taylor and Francis Ltd. Pp. 163-181.

Lemercier, V., X. Millot, E. Ansoborlo, F. Menetrier, A. Flury-Herard, C. Rousselle, and J. M. Scherrmann. 2003. Study of uranium transfer across the blood-brain barrier. *Radiation Protection Dosimetry* 105(1-4):243-245.

Lestaevel, P., C. Bussy, F. Paquet, B. Dhieux, D. Clarencon, P. Houpert, and P. Gourmelon. 2005a. Changes in sleep-wake cycle after chronic exposure to uranium in rats. *Neurotoxicology and Teratology* 27(6):835-840.

Lestaevel, P., P. Houpert, C. Bussy, B. Dhieux, P. Gourmelon, and F. Paquet. 2005b. The brain is a target organ after acute exposure to depleted uranium. *Toxicology* 212(2-3):219-226.

Lin, R. H., W. M. Fu, and S. Y. Lin-Shiau. 1988. Presynaptic action of uranyl nitrate on the phrenic nerve-diaphragm preparation of the mouse. *Neuropharmacology* 27(8):857-863.

Llobet, J. M., J. J. Sirvent, A. Ortega, and J. L. Domingo. 1991. Influence of chronic exposure to uranium on male reproduction in mice. *Fundamental and Applied Toxicology* 164:821-829.

Lopez, R., P. L. Diaz Sylvester, A. M. Ubios, and R. L. Cabrini. 2000. Percutaneous toxicity of uranyl nitrate: Its effect in terms of exposure area and time. *Health Physics* 78(4):434-437.

Malard, V., O. Prat, E. Darrouzet, F. Berenguer, N. Sage, and E. Quemeneur. 2005. Proteomic analysis of the response of human lung cells to uranium. *Proteomics* 5(17):4568-4580.

McClain, D. E., and A. C. Miller. 2007. Depleted uranium biological effects: Introduction and early in vitro and in vivo studies. In *Depleted uranium: Properties, uses, and health consequences*. Boca Raton, FL: CRC Press. Pp. 1-20.

McDiarmid, M. 2007 (unpublished). *Depleted uranium (DU) follow-up program update. Presentation to the IOM committee on Gulf War and health: Depleted uranium*. College Park, MD: University of Maryland. Available: National Academies' Public Access Records Office (publicac@nas.edu).

McDiarmid, M. A., K. Squibb, S. Engelhardt, M. Oliver, P. Gucer, P. D. Wilson, R. Kane, M. Kabat, B. Kaup, L. Anderson, D. Hoover, L. Brown, and D. Jacobson-Kram. 2001. Surveillance of depleted uranium exposed gulf war veterans: Health effects observed in an enlarged "Friendly fire" Cohort. *Journal of Occupational and Environmental Medicine* 43(12):991-1000.

McDiarmid, M. A., S. Engelhardt, M. Oliver, P. Gucer, P. D. Wilson, R. Kane, M. Kabat, B. Kaup, L. Anderson, D. Hoover, L. Brown, B. Handwerger, R. J. Albertini, D. Jacobson-Kram, C. D. Thorne, and K. S. Squibb. 2004. Health effects of depleted uranium on exposed gulf war veterans: A 10-year follow-up. *Journal of Toxicology and Environmental Health A* 67(4):277-296.

Milacic, S., D. Petrovic, D. Jovicic, R. Kovacevic, and J. Simic. 2004. Examination of the health status of populations from depleted-uranium-contaminated regions. *Environmental Research* 95(1):2-10.

Miller, A. C., W. F. Blakely, D. Livengood, T. Whittaker, J. Xu, J.W. Ejnik, M. M. Hamilton, E. Parlette, T. S. John, H. M. Gerstenberg, and H. Hsu. 1998a. Transformation of human osteoblast cells to the tumorigenic phenotype by depleted uranium-uranyl chloride. *Environmental Health Perspectives* 106(8):465-471.

Miller, A. C., A. F. Fuciarelli, W. E. Jackson, E. J. Ejnik, C. Emond, S. Strocko, J. Hogan, N. Page, and T. Pellmar. 1998b. Urinary and serum mutagenicity studies with rats implanted with depleted uranium or tantalum pellets. *Mutagenesis* 13(6):643-648.

Miller, A. C., J. Xu, M. Stewart, and D. McClain. 2001. Suppression of depleted uranium-induced neoplastic transformation of human cells by the phenyl fatty acid, phenyl acetate: Chemoprevention by targeting the P21RAS protein pathway. *Radiation Research* 155(1 Pt 2):163-170.

Miller, A. C., M. Stewart, K. Brooks, L. Shi, and N. Page. 2002a. Depleted uranium-catalyzed oxidative DNA damage: Absence of significant alpha particle decay. *Journal of Inorganic Biochemistry* 91(1):246-252.

Miller, A. C., J. Xu, M. Stewart, K. Brooks, S. Hodge, L. Shi, N. Page, and D. McClain. 2002b. Observation of radiation-specific damage in human cells exposed to depleted uranium: Dicentric frequency and neoplastic transformation as endpoints. *Radiation Protection Dosimetry* 99(1-4):275-278.

Miller, A. C., J. Xu, M. Stewart, P. G. Prasanna, and N. Page. 2002c. Potential late health effects of depleted uranium and tungsten used in armor-piercing munitions: Comparison of neoplastic transformation and genotoxicity with the known carcinogen nickel. *Military Medicine* 167(2 Suppl):120-122.

Miller, A. C., K. Brooks, J. Smith, and N. Page. 2004. Effect of the militarily-relevant heavy metals, depleted uranium and heavy metal tungsten-alloy on gene expression in human liver carcinoma cells (HepG2). *Molecular and Cellular Biochemistry* 255(1-2):247-256.

Miller, A. C., C. Bonait-Pellie, R. F. Merlot, J. Michel, M. Stewart, and P. D. Lison. 2005. Leukemic transformation of hematopoietic cells in mice internally exposed to depleted uranium. *Molecular and Cellular Biochemistry* 279(1-2):97-104.

Miller, A. C., M. Stewart, R. Rivas, S. Marino, G. Randers-Pehrson, and L. Shi. 2007. Observation of radiation-specific damage in cells exposed to depleted uranium: Hprt gene mutation frequency. *Radiation Measurements* 42:1029-1032.

Mitchel, R. E. J., J. S. Jackson, and B. Heinmiller. 1999. Inhaled uranium ore dust and lung cancer risk in rats. *Health Physics* 76(2):145-155.

Monleau, M., C. Bussy, P. Lestaevel, P. Houpert, F. Paquet, and V. Chazel. 2005. Bioaccumulation and behavioural effects of depleted uranium in rats exposed to repeated inhalations. *Neuroscience Letters* 390(1):31-36.

Monleau, M., E. Blanchardon, M. Claraz, F. Paquet, and V. Chazel. 2006a. The effect of repeated inhalation on the distribution of uranium in rats. *Journal of Toxicology and Environmental Health A* 69(17):1629-1649.

Monleau, M., M. De Meo, S. Frelon, F. Paquet, M. Donnadieu-Claraz, G. Dumenil, and V. Chazel. 2006b. Distribution and genotoxic effects after successive exposure to different uranium oxide particles inhaled by rats. *Inhalation Toxicology* 18(11):885-894.

Monleau, M., M. De Meo, F. Paquet, V. Chazel, G. Dumenil, and M. Donnadieu-Claraz. 2006c. Genotoxic and inflammatory effects of depleted uranium particles inhaled by rats. *Toxicological Sciences* 89(1):287-295.

Morris, K. J., C. L. Barker, A. L. Batchelor, and P. Khanna. 1992. Dosimetric implications of pulmonary macrophage clusters observed within lungs of rats that have inhaled enriched UoinF 2 particles. *Environmental Health Perspectives* 97:201-208.

Muller, D., P. Houpert, J. Cambar, and M. H. Henge-Napoli. 2006. Role of the sodium-dependent phosphate co-transporters and of the phosphate complexes of uranyl in the cytotoxicity of uranium in LLC-PK1 cells. *Toxicology and Applied Pharmacology* 214(2):166-177.

Neuman, W. F., and M. W. Neuman. 1948. The deposition of uranium in bone: II radioautographic studies. *Journal of Biological Chemistry* 175:711-714.

Neuman, W. F., R. W. Fleming, A. L. Dounce, A. B. Carlson, J. O'Leary, and B. J. Mulryan. 1948a. Distribution and secretion of injected uranium. *Journal of Biological Chemistry* 173:737-748.

Neuman, W. F., M. W. Neuman, and B. J. Mulryan. 1948b. The deposition of uranium in bone: I animal studies. *Journal of Biological Chemistry* 175:705-709.

NRC (National Research Council). 1991. *Animals as sentinels of environmental health hazards.* Washington, DC: National Academy Press.

———. 2008. *Review of toxicologic and radiologic risks to military personnel from exposure to depleted uranium during and after combat.* Washington, DC: The National Academies Press.

Pappo, J., and T. H. Ermak. 1989. Uptake and translocation of fluorescent latex particles by rabbit Peyer's Patch follicle epithelium: A quantitative model for M cell uptake. *Clinical and Experimental Immunology* 76(1):144-148.

Paquet, F., P. Houpert, E. Blanchardon, O. Delissen, C. Maubert, B. Dhieux, A. M. Moreels, S. Frelon, and P. Gourmelon. 2006. Accumulation and distribution of uranium in rats after chronic exposure by ingestion. *Health Physics* 90(2):139-147.

Paternain, J. L., J. L. Domingo, A. Ortega, and J. M. Llobet. 1989. The effects of uranium on reproduction, gestation, and postnatal survival in mice. *Ecotoxicology and Environmental Safety* 17(3):291-296.

Pellmar, T. C. 1997. *Health risk assessment of embedded depleted uranium: Behavior, physiology and histology—6 month time point*. Bethesda, MD: AFRRI Special Report 97-4.

Pellmar, T. C., A. F. Fuciarelli, J. W. Ejnik, M. Hamilton, J. Hogan, S. Strocko, C. Emond, H. M. Mottaz, and M. R. Landauer. 1999a. Distribution of uranium in rats implanted with depleted uranium pellets. *Toxicological Sciences* 49(1):29-39.

Pellmar, T. C., D. O. Keyser, C. Emery, and J. B. Hogan. 1999b. Electrophysiological changes in hippocampal slices isolated from rats embedded with depleted uranium fragments. *Neurotoxicology* 20(5):785-792.

Periyakaruppan, A., F. Kumar, S. Sarkar, C. S. Sharma, and G. T. Ramesh. 2007. Uranium induces oxidative stress in lung epithelial cells. *Archives of Toxicology* 81(6):389-395.

Petitot, F., A. M. Moreels, and F. Paquet. 2004. In vitro evaluation of percutaneous diffusion of uranyl nitrate through intact or excoriated skin of rat and pig. *Canadian Journal of Physiology and Pharmacology* 82(2):133-139.

Petitot, F., S. Frelon, A. M. Moreels, M. Claraz, O. Delissen, E. Tourlonias, B. Dhieux, C. Maubert, and F. Paquet. 2007. Incorporation and distribution of uranium in rats after a contamination on intact or wounded skin. *Health Physics* 92(5):464-474.

Powell, J. J., C. C. Ainley, R. S. Harvey, I. M. Mason, M. D. Kendall, E. A. Sankey, A. P. Dhillon, and R. P. Thompson. 1996. Characterisation of inorganic microparticles in pigment cells of human gut associated lymphoid tissue. *Gut* 38(3):390-395.

Pujadas Bigi, M. M., and A. M. Ubios. 2007. Catch-up of delayed tooth eruption associated with uranium intoxication. *Health Physics* 92(4):345-348.

Raymond-Whish, S., L. P. Mayer, T. O'Neal, A. Martinez, M. A. Sellers, P. J. Christian, S. L. Marion, C. Begay, C. R. Propper, P. B. Hoyer, and C. A. Dyer. 2007. Drinking water with uranium below the U.S. EPA water standard causes estrogen receptor-dependent responses in female mice. *Environmental Health Perspectives* 115(12):1711-1716.

Roberts, E. 1949. Uranyl nitrate. In *Pharmacology and toxicology of uranium compounds*, edited by C. Voegtlin and H. C. Hodge. New York: McGraw-Hill. Pp. 561-585.

Souidi, M., Y. Gueguen, C. Linard, N. Dudoignon, S. Grison, C. Baudelin, C. Marquette, P. Gourmelon, J. Aigueperse, and I. Dublineau. 2005. In vivo effects of chronic contamination with depleted uranium on CYP3A and associated nuclear receptors PXR and car in the rat. *Toxicology* 214(1-2):113-122.

Stearns, D. M., M. Yazzie, A. S. Bradley, V. H. Coryell, J. T. Shelley, A. Ashby, C. S. Asplund, and R. C. Lantz. 2005. Uranyl acetate induces hprt mutations and uranium-DNA adducts in Chinese hamster ovary EM9 cells. *Mutagenesis* 20(6):417-423.

Stevens, W., F. W. Bruenger, D. R. Atherton, J. M. Smith, and G. N. Taylor. 1980. The distribution and retention of hexavalent 233U in the beagle. *Radiation Research* 83(1):109-126.

Stopps, G. J., and M. Todd. 1982. *The chemical toxicity of uranium with special reference to effects on the kidney and the use of urine for biological monitoring*. Ottawa, Canada: Atomic Energy Control Board.

Thiebault, C., M. Carriere, S. Milgram, A. Simon, L. Avoscan, and B. Gouget. 2007. Uranium induces apoptosis and is genotoxic to normal rat kidney (NRK-52E) proximal cells. *Toxicological Sciences* 98(2):479-487.

Thorne, M. C. 2003. Estimation of animal transfer factors for radioactive isotopes of iodine, technetium, selenium and uranium. *Journal of Environmental Radioactivity* 70(1-2):3-20.

Thun, M. J., D. B. Baker, K. Steenland, A. B. Smith, W. Halperin, and T. Berl. 1985. Renal toxicity in uranium mill workers. *Scandinavian Journal of Work, Environment and Health* 11(2):83-90.

Ting, B. G., D. C. Paschal, J. M. Jarrett, J. L. Pirkle, R. J. Jackson, E. J. Sampson, D. T. Miller, and S. P. Caudill. 1999. Uranium and thorium in urine of United States residents: Reference range concentrations. *Environmental Research* 81(1):45-51.

Tissandie, E., Y. Gueguen, J. M. Lobaccaro, F. Paquet, J. Aigueperse, and M. Souidi. 2006. Effects of depleted uranium after short-term exposure on vitamin D metabolism in rat. *Archives of Toxicology* 80(8):473-480.

Tissandie, E., Y. Gueguen, J. M. Lobaccaro, L. Grandcolas, P. Voisin, J. Aigueperse, P. Gourmelon, and M. Souidi. 2007. In vivo effects of chronic contamination with depleted uranium on vitamin D3 metabolism in rat. *Biochimica et Biophysica Acta* 1770(2):266-272.

Tracy, B. L., J. M. Quinn, J. Lahey, A. P. Gilman, K. Mancuso, A. P. Yagminas, and D. C. Villeneuve. 1992. Absorption and retention of uranium from drinking water by rats and rabbits. *Health Physics* 62(1):65-73.

Tymen, H., P. Gerasimo, and D. Hoffschir. 2000. Contamination and decontamination of rat and human skin with plutonium and uranium, studied with a Franz's chamber. *International Journal of Radiation Biology* 76(10):1417-1424.

Ubios, A. M., M. B. Guglielmotti, and R. L. Cabrini. 1990. Ethane 1-hydroxy-1, 1-diphosphonate (EHDP) counteracts the inhibitory effect of uranyl nitrate on bone formation. *Archives of Environmental Health* 45(6):374-377.

Ubios, A. M., E. M. Braun, and R. L. Cabrini. 1998. Effect of biphosphonates on abnormal mandibular growth of rats intoxicated with uranium. *Health Physics* 75(6):610-613.

USACHPPM (U.S. Army Center for Health Promotion and Preventive Medicine). 2004. *Capstone report: Depleted uranium aerosol doses and risks: Summary of U.S. Assessments*. Fort Belvoir, VA: US Army Heavy Metals Office, Chemical and Biological Defense Information Analysis Center.

Voegtlin, C., and H. C. Hodge. 1949. *Pharmacology and toxicity of uranium compounds*. Vol. 1. New York: McGraw-Hill.

Wan, B., J. T. Fleming, T. W. Schultz, and G. S. Sayler. 2006. In vitro immune toxicity of depleted uranium: Effects on murine macrophages, CD4+ T cells, and gene expression profiles. *Environmental Health Perspectives* 114(1):85-91.

Wise, S. S., W. D. Thompson, A. M. Aboueissa, M. D. Mason, and J. P. Wise, Sr. 2007. Particulate depleted uranium is cytotoxic and clastogenic to human lung cells. *Chemical Research in Toxicology* 20(5):815-820.

Yazzie, M., S. L. Gamble, E. R. Civitello, and D. M. Stearns. 2003. Uranyl acetate causes DNA single strand breaks in vitro in the presence of ascorbate (vitamin C). *Chemical Research in Toxicology* 16(4):524-530.

4

Methodology

This chapter describes the committee's approach to its task. The committee's initial step was to conduct a comprehensive search of the scientific literature to identify studies of long-term health outcomes in humans that might be associated with exposure to depleted uranium. The committee developed criteria for evaluating the relevance and quality of studies. The selected studies constituted the primary evidence from which the committee drew conclusions about the relationships between uranium and specific long-term health outcomes. The committee then ranked the strength of the relationships by using the five-category system presented at the end of the chapter.

INFORMATION-GATHERING STRATEGY

The committee used a multistep process to identify scientific studies of the long-term health outcomes of exposure to uranium compounds, including depleted uranium. Twelve data sources—including PubMed, the National Technical Information Service, and Toxicology Literature Online (TOXLINE)—were searched for the key phrase *depleted uranium* and the Medical Subject Heading (MeSH) terms *uranium* and *uranium compounds*. Uranium-related studies identified by the committee that wrote *Gulf War and Health, Volume 1: Depleted Uranium, Pyridostigmine Bromide, Sarin, Vaccines* (IOM, 2000b; hereafter referred to as *Volume 1*) were added to the current committee's reference database. Additional studies were identified from the reference lists of technical reports, books, and other documents.

The searches generated about 3,500 titles and abstracts, which were examined to identify articles that appeared to be relevant to the committee's task, that is, articles on health outcomes of exposure to uranium (including both natural and depleted uranium). Examples of types of articles that were excluded during this step are environmental studies (that is, effects on wildlife), engineering studies of nuclear reactors, studies of the treatment and disposal of uranium, studies of naturally occurring uranium concentrations in various locations, bioremediation studies, and studies of nuclear-plant safety. After examination of the titles and abstracts, about 1,000 articles remained in the committee's reference database. The data sources listed above were searched on a monthly basis through December 2007 and relevant articles were added to the reference database.

To gather further information, the committee held a public meeting on June 28, 2007, in Washington, DC. The topics discussed are shown in Box 4-1. Four speakers gave presentations related to health effects of exposure to depleted uranium and uranium in human populations, including the veteran population. Another presenter discussed toxicologic studies of uranium. After the formal presentations, the floor was opened to members of the public who wished to make comments.

BOX 4-1
Open-Session Presentations, June 28, 2007

Depleted Uranium Exposure and Health Effects in Gulf War Veterans	Melissa McDiarmid, Department of Veterans Affairs Depleted Uranium Follow-Up Program
Outcomes of the UK Ministry of Defence–Sponsored Depleted Uranium Research Programme	Nicholas Priest, Atomic Energy of Canada Limited, Canada (formerly with Middlesex University, UK)
Depleted Uranium and Veterans' Health: A Flawed Testing Process and an Undersized, Politicized Study Limit Evaluation of Exposures and Effects	Dan Fahey, Board Member, Veterans for Common Sense, and PhD candidate, University of California, Berkeley
Current NIOSH Research on Uranium-Exposed Workers	Mary Schubauer-Berigan, National Institute for Occupational Safety and Health
Toxic and Radiologic Effects of Uranium: Animal Studies	Fletcher Hahn, Lovelace Respiratory Research Institute
Department of Defense Health Databases	Kenneth Cox, Department of Defense

PRINCIPAL OBJECTIVES OF EPIDEMIOLOGIC STUDIES

Epidemiologic studies examine the relationship between exposures to agents of interest—uranium in this case—in a human population and the health outcomes seen in the population. The challenge of epidemiologic studies is to control for risk factors that are related to the exposures and health outcomes of interest by using study designs and statistical techniques that control for bias and confounding. Such studies can be used to generate hypotheses for future study or to test hypotheses posed by investigators.

A principal objective of epidemiology is to understand whether exposures to specific agents are associated with disease or other health outcomes and to evaluate whether such associations are potentially causal. Although they are often used synonymously by the general public, the terms *association* and *causation* have distinct meanings (Alpert and Goldberg, 2007).

Epidemiologic studies can establish statistical associations between exposures and health effects, and associations are generally expressed by using relative risks or odds ratios. To conclude that an association exists, it is necessary for an exposure to be followed by a health effect more frequently than would be expected by chance alone. Furthermore, confidence in an association rises when it is consistently observed in several studies. However, the results of separate studies are sometimes conflicting. It is possible to attribute discordant study results to differences in such characteristics as soundness of study design, quality of execution, and the influence of different forms of bias. Studies that result in a tight confidence interval around a statistically significant relative risk of association constitute stronger evidence of an effect. When the measure of association does not show a statistically significant effect, it is important to consider the size of the sample and whether the study had the power to detect an effect of a given size. Epidemiologic study designs differ in their ability to provide valid estimates of an association (Ellwood, 1998). Cross-sectional studies generally provide a lower level of evidence than cohort and case-control studies.

Determining whether a given statistical association rises to the level of causation requires inference (Hill, 1965). As discussed by the International Agency for Research on Cancer (IARC) in the preamble of its monographs on evaluating cancer risks (for example, IARC, 2004), a strong association is demonstrated by repeated observations in a number of studies, an increased risk of disease with increasing exposure or a decline in risk after cessation of exposure, and specificity of an effect. Those characteristics all strengthen the likelihood that an association seen in epidemiologic studies is a causal effect. Inferences from epidemiologic studies, however, are often limited to population or ecologic associations because of a lack of information on individual exposures. Exposures are rarely, if ever, controlled in epidemiologic studies, and there is usually large uncertainty in the assessment of exposure. To assess whether explanations other than causality are responsible for an observed association, one must bring together evidence from different studies and apply well-established criteria, which have been refined over more than a century (Hill, 1965; Susser, 1973, 1977, 1988, 1991; Evans, 1976; Wegman et al., 1997). For a review of those criteria, see the 2004 report of the US Surgeon General (2004).

In examining the available epidemiologic studies, the committee addressed the question, "Does the available evidence support a causal relationship or an association between exposure to uranium and a health effect?" Even a causal relationship between exposure to uranium and a specific health effect would not mean that uranium invariably results in the health effect or that all cases of the effect are the result of uranium exposure. Such complete correspondence between exposure and disease is the exception in large populations (IOM, 1994b). The committee evaluated the data and based its conclusions on the strength and coherence of the data in the selected epidemiologic studies that met its inclusion criteria.

FACTORS INFLUENCING THE RELEVANCE AND QUALITY OF STUDIES

The committee considered several important issues in its evaluation of the epidemiologic studies and assessment of evidence on uranium-processing workers and civilian and deployed populations exposed to natural and depleted uranium. The discussion builds on the topics covered in *Volume 1* inasmuch as many of the methodologic issues are common to the old and new studies. Like the past committee, the present committee considered a number of factors in its evaluation, including measurement of exposure, assessment of outcome, and relevance of the study population to veterans. As a result, the committee has also outlined the common limitations in the studies on which it based its opinions. The basic limitations are a lack of representativeness or applicability, potential selection bias, incomplete control for potential confounders, and exposure and outcome misclassification. Limitations peculiar to specific types of studies are included in the discussion of those studies (see Chapter 7).

Study Populations

Relevance to Veteran Populations

One of the most important potential limitations of existing studies is the lack of applicability to veteran populations. With the exception of the studies on deployed populations, many of the study samples differed from the Gulf War and Operation Iraqi Freedom troops with respect to characteristics and risk factors. The uranium-processing workers, for example, were mostly white men; few analyses examined health effects in minority groups or women. Some of the residential studies included people of all ages (such as children and the elderly) living in a particular region with little regard to demographics, lifestyle, or other risk factors.

Comparison-Group Issues

Many of the cohort studies of occupationally exposed workers described in Chapter 7 compared death rates in workers with death rates in the US population (or the population of the counties or states in which uranium workers lived). Those studies reported the standardized mortality ratio (SMR) because it is the principal means used in occupational studies to express the death rate in workers relative to that in people not exposed to the agent being studied. A statistically significantly increased SMR (greater than 100 with 95% confidence limits that do not include 100) indicates the possibility of an association between an exposure and a disease. Replication of such a finding strengthens the evidence of an association.

In the case of the occupational cohorts, the results might be skewed because of the "healthy-worker effect." That uranium processors were routinely subject to health examinations before and during employment means that this cohort was generally healthier than the population at large. Thus, one would expect mortality and morbidity in workers to be lower than those in the general population as reflected in study results regardless of exposure; this is known as the healthy-worker effect. It should be noted that studies of military populations are often subject to the same bias (the "healthy-warrior effect") and this type of selection bias may result in underestimation of the association of the exposure and an outcome.

The best way to avoid the healthy-worker bias is to use an internal comparison group of employed people who did not receive the exposure of interest. However, even internal comparison groups can be subject to the bias to the extent that less healthy workers may not stay in more physically demanding jobs or jobs—like uranium milling—that may involve greater exposure to chemical agents. Studies that use internal comparison groups are generally more valid than studies that use external comparison groups (such as the US population or the population of a region) but are also subject to bias. It may be difficult to draw conclusions from studies that directly compare the SMRs of groups of workers that experienced different levels of radiation exposure, because the influence of confounding variables may differ between the groups.

Exposure Assessment

Methods for measuring exposure varied among the three study types—studies of uranium processors and of civilian and deployed populations. Studies of occupational exposure on which the committee relied heavily to evaluate the effect of uranium on disease used several methods and models to assess exposure, including direct measurement of individual exposure through estimates of internal and external radiation dose, the use of work histories to estimate cumulative exposure, and classification of workers by maximum exposure.

Direct Measurement in Individual Workers

The preferred method for an occupational study (or for any study) is to measure exposure of each worker directly. Radiation film badges give a measure of cumulative exposure but respond only to external radiation, which is of greater concern for exposure to enriched uranium than to natural or depleted uranium. Measuring the internal dose of radiation is more difficult. The best method is mathematical modeling to infer the lung dose of uranium from measurements of uranium in the urine or ambient dust.

However, the direct-measurement approach requires that a company monitor each individual worker for radiation exposure and keep thorough, accurate records. In many of the occupational retrospective cohort studies, the authors found that measurements of exposure in individual workers were either unavailable or unreliable. In some cases, records were incomplete, so measurements missing for many workers were estimated from earlier periods or neighboring worksites. In other cases, the only measurements were of urinary uranium excretion. The body excretes uranium rapidly, so urinary uranium is a measure only of exposure in the preceding several days, not over the extended work period.

Using Work History to Model Cumulative Exposure

Several researchers approximated individual exposure by modeling cumulative exposure on the basis of a worker's job history in the plant and the level of exposure in each worksite. They measured uranium exposure in various worksites in the processing plant, using measures of urinary uranium or uranium in ambient dust. That information was used to model the cumulative lung dose per unit time in the worksite. They then used plant employment records to determine the amount of time that each worker spent in each job. By totaling each worker's cumulative exposure in various worksites over the course of the worker's period of employment, they estimated the worker's total exposure.

The modeling approach in effect assigns to each worker the average exposure in each worksite. Compared with individual direct measurement, this approach loses specific information because workers in a given site may vary in their exposure. Any approach that randomly misclassifies individual workers' exposure levels while accurately tracking their health outcomes will result in muted estimates of association between exposure and outcome. That biases a study toward failing to detect an association between exposure and a health outcome even if one exists.

Classifying Workers by Maximum Exposure

This approach measures average exposure in each worksite, as described in the preceding section, and classifies worksites into a relatively small number of groups according to the level of exposure. However, instead of estimating cumulative exposure over all worksites, this method uses as the exposure level for a worker the highest exposure level among all sites to which the worker was assigned for a minimum period (usually 1 month).

This approach of exposure modeling is even cruder because it reduces the variation among workers' exposure levels in two ways. First, it assumes that an employee spent his or her entire period of employment in one group of worksites, whereas the worker may have spent time in sites that varied considerably in exposure levels. Second, it combines sites that may vary considerably in their level of exposure. For those reasons, this approach is especially prone to false-negative results (that is, failing to detect a dose-response relationship). However, the effect of the shortcomings of this approach is unknown because none of the studies estimated the probability of false-negative results.

Self-Reporting in Exposure Assessment

Self-reporting of exposure is a potential limitation. People's ability to recall details of exposure over a period of years accurately can vary widely and is likely to be small. In addition, recall can be influenced (that is, biased) by whether a person has experienced an adverse health outcome. Relying on memory can result in imprecise and even invalid assessments of exposure.

Other Methods of Estimating Exposures

A study that does not classify workers according to exposure cannot use workers with low exposure as an internal control when estimating the health effects of high exposure. Such a study must use the US population or the population of the region in which the plant is sited as the external control group. In that approach, the healthy-worker effect is more likely to distort estimates of the effect of exposure on health outcomes, generally biasing results toward lower risk among the exposed.

Many of the studies were limited by potential exposure misclassification. Exposure can often be difficult to measure, particularly in settings where participants were exposed to a variety of compounds. For example, the uranium-processing workers received a variety of chemical and radiologic coexposures that are impossible to separate, so it is impossible to evaluate which exposure resulted in the outcome of interest. In addition, dose plays a crucial role in risk, and in many cases doses could not be determined. The studies also suffer from the lack of technical precision that is available today, so both overall exposure and dosage could be only imprecisely estimated on the basis of such surrogate factors as job classification or assignment.

In the residential studies, except the drinking-water studies, exposure was generally not measured at the individual level, and exposure assessment was based on geographic proximity modeling. In such studies, dose-response relationships cannot be determined. The same is true of the studies of deployed personnel; in war situations (particularly combat, in which depleted-uranium exposure is most likely), environmental monitoring is not feasible.

Outcome Assessment

Biologic Plausibility

Biologic plausibility reflects knowledge of the biologic mechanism by which an agent can lead to a health outcome. That knowledge comes through mechanism-of-action or other studies in pharmacology, toxicology, microbiology, physiology, and other fields—typically in studies of animals. Biologic plausibility is often difficult to establish or may not be known when an association is first documented. The committee considered such factors as evidence from animal and human studies that exposure to an agent is associated with diseases known to have biologic mechanisms similar to that of the disease in question, evidence that some outcomes are commonly associated with occupational or environmental exposures, and knowledge of routes of exposure, storage in the body, and excretion that suggests that a disease is more likely to occur in some organs than in others. The extent to which the data are consistent with a biologically plausible mechanism influences the weight attached to the results of a study, as does an indication that the mechanism is similar in the animals under study and humans.

Biomarkers

A biomarker is a molecular or cellular indicator of exposure, effect, or susceptibility. More specifically, a biomarker of effect is defined as a "measurable biochemical, physiologic, behavioral, or other alteration in an organism that, depending on the magnitude, can be recognized as associated with an established or possible health impairment or disease" (NRC, 2006). Biomarkers of effect can include biochemical, cellular, and physiologic indicators of disease. Numerous studies evaluated by the committee incorporated biomarkers of effect to evaluate health outcomes related to uranium exposure, including biomarkers to evaluate cell toxicity and renal dysfunction.

Adequate Followup Period

To strengthen the evidence of a true cause-effect association (particularly for some health outcomes, such as most cancers), the followup period should allow sufficient time after exposure for the health outcome to occur in the population of concern. There are several time-related factors. Biologic latency of cancer is a factor in the delay between exposure to a putative carcinogen and the appearance of cancer. For most cancers, the lag between exposure and diagnosis is at least 10 years; however, there are exceptions, such as leukemia. Eliminating study participants who died from cancer that occurred within 10 years of exposure should increase the SMR if there is a true association between exposure to the agent and the cancer. Conversely, the case for an association is much weaker when the death rate relative to that in the US population is the same whether or not the author considered the early cancer deaths.

Specificity of Outcome

The study had to specify a distinct outcome rather than a nonspecific group of health outcomes. Lack of specificity occurs primarily in mortality studies that examine all-cause mortality (such as deaths from all types of cancer) as opposed to cause-specific mortality (such as deaths from lung cancer). All-cause mortality studies were excluded unless they analyzed specific health outcomes.

Adverse Clinical Outcomes

After reviewing the approximately 1,000 articles in the reference database, the committee focused on a number of relevant health outcomes on which to draw conclusions (see Chapters 6 and 8). The selected health outcomes are 10 types of cancer and several nonmalignant diseases or conditions. The types of cancer are lung cancer, leukemia, lymphoma, bone cancer, renal cancer, bladder cancer, brain and other central nervous system cancers, stomach cancer, prostatic cancer, and testicular cancer; the nonmalignant diseases or conditions include renal disease, respiratory disease, neurologic disease, and reproductive and developmental effects. With the exception of prostatic and testicular cancers, the health outcomes were selected by the committee because there are plausible mechanisms of action (for example, lung cancer and respiratory disease were selected because inhaled insoluble uranium oxides lodge in the lung). Prostatic cancer is the most frequently diagnosed cancer in men in the United States, and any slight increase in risk could result in large numbers of cases and deaths. Testicular cancer, the most common cancer in young men, is of special interest to Gulf War veterans, and some recent studies of veterans suggested a higher but nonsignificant risk in them than in their nondeployed counterparts (IOM, 2006).

Considerations in Statistical Inference

Tests of Association

Studies of a possible relationship between an exposure of interest and an outcome typically report statistical tests of association. Those tests assess whether the data are consistent with the claim of an association between exposure and outcome. The association is commonly expressed in terms of null and alternative hypotheses. The null is chosen to be consistent with the "status quo"—statistical independence or no association between exposure and outcome. The alternative is chosen to represent the opposite point of view: that an association exists between exposure and outcome (that is, they are not independent). A summary statistic, called a test statistic, is calculated that gauges how well the data "match" the null hypothesis. In general, small values of the test statistic reflect consistency with the null, and large values consistency with the alternative. The magnitude of the test statistic is compared with its expected size under the null hypothesis. The difference between the observed value of the statistic and its expected value under the null is evaluated while taking into consideration such factors as the size of the sample and variability of the measurements.

The p Value

In reporting the results of a statistical test of association, researchers report a p value, or the probability of observing a test statistic as large as or larger than (in absolute value) that obtained from the sample if the null hypothesis is true. Small p values therefore indicate that the probability of observing a result as extreme as or more extreme than that obtained in the study is very unlikely if the null is true. By convention, most researchers use a p value of 0.05 as the threshold value for rejecting the null hypothesis. Therefore, if researchers observe $p < 0.05$, they state that a result is "statistically significant"; if the p value exceeds 5%, they state that the result is nonsignificant.

Type I and Type II Error and Power

It is possible to make two types of errors in conducting a statistical test of association. First, the null hypothesis might be rejected when it is true, simply because of chance variation. That is called a type I error, or α. The second type of error is failure to reject the null hypothesis when the alternative is true. That is called a type II error. One minus the type II error, or the probability of rejecting the null when the alternative is true, is called the power of a test. In general, researchers want both error rates to be low. In practice, the type I error is usually set to an acceptable level (usually 5%, as indicated above), and a study is designed to obtain a suitably large value for the power. Power is a function of the size of the study sample, the duration of followup, and the strength of the exposure effect. Longer followup will also allow examination of a range of latent periods between exposure and diagnosis of disease.

Control of Bias

Bias refers to systematic or nonrandom error. Bias causes an observed value to deviate from the true value. It can weaken an association or generate a spurious association. Because all studies are susceptible to bias, a goal is to minimize bias or to adjust the observed value of an association by using special methods to correct for bias. Two kinds of bias may compromise the results of an investigation: selection bias and information bias.

- *Selection bias* occurs when the participants in a study are not representative of the general population. The study participants differ from nonparticipants in characteristics that cannot be observed, that is, groups differ in measured or unmeasured baseline characteristics because of how participants were selected or assigned.
- *Information bias* results from the manner in which data are collected and can result in measurement errors, imprecise measurement, and misdiagnosis. Those types of errors may be uniform in an entire study population or may affect some parts of the population more than others. Bias may result from misclassification of study subjects with respect to the outcome variable. Other common sources of information bias are the inability of study subjects to recall accurately the circumstances of their exposure (recall bias) and the likelihood that one group more frequently reports what it remembers than another group (reporting bias). Information bias is especially harmful in interpreting study results when it affects one comparison group more than another.

Coexposures

Confounders

Many of the studies reviewed failed to control for potential confounders. For many of the outcomes of interest, there are several well-known risk factors that were not taken into consideration; these include smoking, diet, other lifestyle factors, and preexisting illness. In some studies, the lack of control was a result of the study design; for example, ecologic studies, such as the residential studies in which exposure is determined solely by geographic proximity to an exposure source, by design cannot take individual-level factors into account. Retrospective cohort studies can only be analyzed on the basis of the data available; often, information on other risk factors was not collected, either because they were not known risk factors at the time or because collection of such information was not routine.

Synergism

Interaction, or synergism, occurs when combined exposure to two or more chemicals is more likely to produce an adverse health outcome than exposure to the chemicals individually. Epidemiologic studies are typically unable to partition data on exposures to multiple chemicals quantitatively and even less likely to be able to attribute health outcomes related to combined exposures. As discussed previously, many of the studies evaluated by the committee, including those with an extensive exposure-assessment component (that is, occupational studies), might not have been able to account for the numerous chemical or radiologic coexposures. Although the committee was not charged with evaluating health effects related to combined exposures to chemicals, it acknowledges the possibility that such exposures are common, particularly in occupational settings.

EPIDEMIOLOGIC-STUDY DESIGNS

The major types of epidemiologic studies evaluated by the committee are cohort, case-control, cross-sectional, ecologic studies, and case reports and case series.

Cohort Studies

A cohort, or longitudinal, study follows a defined group, or cohort, over time. It can test hypotheses about whether an exposure to a specific agent is related to the development of a health effect and can examine multiple health effects that may be associated with exposure to a given agent. A cohort study starts by classifying study participants according to whether they have been exposed to the agent under study, in this case uranium. A cohort study compares health effects in people who have been exposed with those in people who have not been exposed. Such a comparison can be used to estimate a risk difference or a relative risk, two statistics that measure association. The risk difference is the rate of disease or health effect in exposed persons minus the rate in nonexposed persons; a value greater than zero implies that an excessive rate of disease is associated with the exposure. The relative risk, or risk ratio, is determined by dividing the rate of the disease in the exposed group by the rate in the nonexposed group; a relative risk greater than 1 suggests a positive association between the agent and the health effect, and the higher the relative risk, the stronger the association.

One major advantage of a cohort study is the ability of the investigator to define the exposure classification of subjects at the beginning of the study. The classification in prospective cohort studies (see below) is not influenced by the presence of a health effect, because the health effect has yet to occur, and this reduces an important source of potential bias known as selection bias. As explained in the next section, on case-control studies, when it is possible to measure a confounding factor,[1] the investigator can apply statistical methods to

[1]A potential confounding factor is a variable that is associated with the health outcome and may affect the results of the study because it is distributed differently in the exposed and nonexposed groups.

minimize its influence on the results. Another advantage of a cohort study is that it is possible to calculate absolute rates of disease incidence.[2] A final advantage, especially over cross-sectional studies (discussed below), is that it may be possible to adjust each subject's followup health status in light of baseline health status so that the person acts as his or her own control instead of defining a group as "disease-free"; this may reduce a source of variation and increase the power to detect effects. The disadvantages of cohort studies are the high costs associated with using a large study population and long periods of followup (especially if the health effect is rare), attrition of study subjects, and delay in obtaining results.

A prospective cohort study selects subjects on the basis of exposure (or lack of it) and follows the cohort to determine the rate at which the health effect develops. A retrospective (or historical) cohort study differs from a prospective study in temporal direction; the investigator traces back in time to classify past exposures in the cohort and then tracks the cohort forward in time to ascertain the rate of the health effect. Retrospective cohort studies often focus on mortality because of the relative ease of determining the vital status of individuals and the availability of death certificates to determine the causes of deaths. Most cohort studies are retrospective.

For comparison purposes, cohort studies often use general population mortality or morbidity rates (age-, sex-, race-, time-, and cause-specific) because it may be difficult to identify a suitable control group of nonexposed people. The observed number of deaths or cases of illness in a group (related to a specific cause, such as lung cancer) is compared with the *expected* number of deaths or cases of illness. The ratio of observed to expected deaths is an SMR. An SMR greater than 1.0 generally suggests an increased risk of deaths in the exposed group.

The major problem in using general population rates for comparison with military-cohort rates is the healthy-warrior effect, which arises when a military population experiences a lower mortality or morbidity rate than the general population, a mixture of healthy and unhealthy people. The military has physical-health criteria that personnel must meet when they enter the military and while they are on active duty.

Case-Control Studies

In a case-control study, subjects (cases) are selected on the basis of having a health effect; controls are selected on the basis of not having the health effect. Cases and controls are asked about their exposures to specific agents. Cases and controls can be matched with regard to such characteristics as age, sex, and socioeconomic status to eliminate those characteristics as causes of observed

---

[2]Incidence is the rate of occurrence of new cases of an illness or disease in a given population during a specified period.

differences, or those variables can be controlled for in the analysis. The odds of exposure to the agent among cases are then compared with the odds of exposure among controls. The comparison generates an odds ratio, which is a statistic that depicts the odds that those exposed to the agent in question will have a health effect relative to the odds that those not exposed will have the health effect. An odds ratio greater than 1 indicates that there is a potential association between exposure to the agent and the health effect; the greater the odds ratio, the stronger the association.

Case-control studies are useful for testing hypotheses about the relationships between exposure to specific agents and a health effect. They are especially useful and efficient for studying the etiology of rare effects. Case-control studies have the advantages of ease, speed, and relatively low cost. They are also valuable for their ability to probe multiple exposures or risk factors. However, case-control studies are vulnerable to several types of bias, such as recall bias, which can dilute or enhance associations between exposure and a health effect. Other problems include identifying representative groups of cases, choosing suitable controls, and collecting comparable information about exposures of cases and controls. Those problems might lead to unidentified confounding variables that differentially influence the selection of cases or control subjects or the detection of exposure. For the reasons discussed above, case-control studies are often the first approach to testing a hypothesis about whether factors contribute to a specific health effect, especially a rare one.

A "nested" case-control study draws cases and controls from a previously defined cohort. Thus, it is said to be nested in a cohort study. Baseline data are collected at the time that the cohort is identified, and this ensures a more uniform set of data on cases and controls. Members of the cohort who are identified as having a health effect serve as cases, and a sample of those who are effect-free serve as controls. Baseline data are used to compare exposure in cases and controls, as in a regular case-control study. Nested case-control studies are efficient with respect to the time and cost needed to reconstruct exposure histories of cases and of only a sample of controls rather than the entire cohort. In addition, because the cases and controls come from the same previously established cohort, concerns about unmeasured confounders and selection bias are decreased.

Cross-Sectional Studies

The main differentiating feature of a cross-sectional study is that exposure information and health-effect information are collected at the same time. The selection of people for the study—unlike selection for cohort and case-control studies—is independent both of the exposure to the agent in question and of health-effect characteristics. Cross-sectional studies seek to uncover potential associations between exposure to a specific agent and development of a health effect. In a cross-sectional study, effect size is measured as relative risk, prevalence[3] ratio, or prevalence odds ratio. The study might compare health-effect or symptom rates in groups with and without exposure to uranium.

Cross-sectional studies are easier and less expensive to perform than cohort studies and can identify the prevalence of health effects and exposures in a defined population. They are useful for generating hypotheses, but they are much less useful for determining cause-effect relationships (Monson, 1990). It might also be difficult to determine the temporal sequence of exposures and symptoms or effect.

Ecologic Studies

Ecologic studies examine the relationship between exposure and disease in groups of people rather than individuals. They can be used as a first step to explore whether an association between exposure and disease exists or to identify avenues of research to investigate etiologic relationships. Ecologic studies require aggregate data on disease and exposure. Data on disease occurrence are commonly derived from incidence and mortality data, and exposure information is often based on an overall index, for example, environmental data, such as air or water quality.

Despite the advantages of being inexpensive and relatively less time-consuming, ecologic studies have numerous methodologic problems. Ecologic studies use population-level data rather than individual-level data to assess the relationship between exposure and outcome, so an observed association cannot be used to draw inferences at the individual level (what would be referred to as the ecologic fallacy). Other methodologic problems include difficulty in controlling for confounders, within-group misclassification, and an inability to detect complicated relationships because of the limited data used in the analysis.

Case Reports and Case Series

A case report is a detailed descriptive study of an individual (case report) or small group (case series) in which an association between a health effect and a specific exposure (in this case uranium) is evaluated on the basis of medical histories and clinical evaluations. Case reports are observational, can be prospective or retrospective, and are most useful in assessing rare diseases or providing an indication that an adverse health effect is related to an exposure. Case-series studies also have a number of limitations: they are vulnerable to bias because the observations are generally uncontrolled and collected in an unsystematic manner, and the small number of observations makes it impossible to generalize the findings to a larger population.

---

[3]Prevalence is the number of cases of an illness or disease in a given population at a specific point or in a specific period.

INCLUSION CRITERIA

After securing the full text of the roughly 1,000 articles mentioned above, the committee had to determine which ones to include in its review. For a study to be included in the committee's review, it had to meet these criteria:

- It had to be a published in a peer-reviewed journal or have undergone an equally rigorous process.
- It had to include details of its methodology.
- It had to include a control or reference group.
- It had to use reasonable methods to control for confounders and minimize selection bias.
- It had to use appropriate assessment of uranium exposure in the population. It had to measure exposure to uranium separately from other exposures; studies that dealt with multiple exposures and did not specifically report uranium exposures were excluded.
- It had to deal with long-term health outcomes.
  - It had to have a followup time adequate to detect an effect.
  - It had to have appropriate outcome assessments and measurements based on the expected biologic mechanisms.
- It had to include a relevant study population. Relevant study populations are
  - Uranium-exposed workers (that is, uranium-processing workers).
  - Military personnel deployed to the Gulf War.
  - People living close to uranium-processing facilities. Uranium exposure of such people may be similar to the "level III" exposure received by military personnel (see Chapter 5).

RATIONALE FOR NOT INCLUDING STUDIES OF URANIUM MINERS

The committee engaged in extensive discussion about the role of epidemiologic studies of uranium miners in shaping its report and its conclusions. The committee that wrote *Volume 1* "examined studies of health effects in uranium miners, but concluded that these studies have limited relevance because the primary disease-causing exposures were not to uranium but to radon decay products" (IOM, 2000b). The current committee also concluded that epidemiologic studies in uranium miners have limited relevance, but for different reasons.

As detailed elsewhere in this report, the committee attempted to separate radiologic and chemical effects. The main health effect of interest in uranium miners is lung cancer, which this committee deems a radiologic effect, although a chemical toxic effect cannot be ruled out. If it is a radiologic effect, it should be theoretically possible to convert exposures to mine-based uranium and battlefield-based depleted uranium into equivalent radiation doses (including corrections for radiation quality) by dose reconstruction and to transfer radiogenic lung-cancer risk estimates from the uranium miners to Gulf War veterans. However, the committee identified four issues related to confounding that substantially limited the usefulness of the uranium-miner studies:

- *Differences in physicochemical properties*. The uranium miners were exposed primarily to radon decay products, whereas the veterans were exposed to aerosolized uranium particles. The two kinds of material probably have differences in physicochemical properties, such as aerodynamic behavior and solubility, that could result in large differences in environmental distribution in different media and thus differences in exposure, in transfer from the environment to humans, and in the amount and pattern of deposition in humans (internal dose).
- *Deposition pattern in the human body, especially the airways*. The uranium miners were exposed primarily to radon decay products that were attached to (adsorbed onto the surface of) other particulate matter in the mine air. That particulate matter was mostly large or "coarse" (greater than 1 μm in diameter). Furthermore, the miners typically used a combination of nose and mouth breathing at a relatively high inspiratory flow rate. Those considerations resulted in aerodynamic properties that favored deposition in large, central airways. In contrast, veterans were exposed to a wider range of particle sizes, and the toxicants of interest (depleted-uranium particles) were not necessarily attached to other particulate matter. Smaller particles would probably penetrate all the way to the periphery of the lung and result in exposure of different types of cells and in the possibility of different types of lung cancer from what was typically the case in uranium miners.
- *Radiological vs chemical mechanisms of toxicity*. If the mechanism of lung-cancer production in the uranium miners is radiologic and if the same and only the same mechanism applies to the veterans, the type of dose reconstruction and risk transfer identified above would be appropriate and useful. However, because depleted-uranium toxicity involves an unknown combination of radiologic and chemical mechanisms, including the possibility of synergism between mechanisms, it is scientifically inappropriate to use the uranium-miner data in this fashion.
- *Coexposures to other agents*. As pointed out in *Volume 1*, the uranium "miners were exposed to other possibly toxic dusts and, potentially, to diesel gas fumes, which might cause cancer and other diseases of the lung" (IOM, 2000b). Veterans were also potentially exposed to a suite of toxic dusts, including diesel gas fumes, and that suite probably differed from that to which the uranium miners were exposed. In addition, smoking, which could have differed in pervasiveness and magnitude between miners and veterans, is an important confounding variable.

In light of those four issues, the committee followed the approach of the first committee: the uranium-miner data were studied but played only a small role in the final assessment of risk.

CATEGORIES OF STRENGTH OF ASSOCIATION

The committee used the evidence in the scientific literature to draw conclusions about exposure to depleted uranium and specific adverse health outcomes. Its conclusions are presented as categories of strength of association.

Origin of the Categories

IARC, part of the World Health Organization, established criteria in 1971 to evaluate the human carcinogenic risk posed by chemicals (IARC, 1998). First published in 1972, IARC's evaluations are scientific, qualitative judgments of ad hoc working groups about the evidence of carcinogenicity provided by the available data. The working groups expressed their qualitative judgments by choosing from among five categories to describe the relative strength of the evidence that a substance or exposure is carcinogenic (IARC, 2006a). That agencies in 57 countries use IARC's published evaluations reflects the wide acceptance of the categorization scheme as it has been updated and applied to about 900 agents, mixtures, and exposures (IARC, 2005, 2006b). In the early 1990s, an Institute of Medicine committee adopted IARC's categories for its evaluation of the adverse health effects of pertussis and rubella vaccines (IOM, 1991). Later committees used the categories, with some modifications, in their evaluations of the safety of childhood vaccines (IOM, 1994a), of the health effects of herbicides used in the Vietnam War (IOM, 1994b, 1996, 1999, 2001, 2003), and of the relationship between exposure to indoor pollutants and asthma (IOM, 2000a). The categories also were adapted and used by the present committee's predecessors, which evaluated the health effects of vaccines given to US troops and of chemical, biologic, and physical exposures that may have occurred during the Gulf War (IOM, 2000a,b, 2004, 2005, 2007).

The five categories used in this report are defined below, and the committee's conclusions are presented in Chapter 8.

Sufficient Evidence of a Causal Relationship

Evidence is sufficient to conclude that a causal relationship exists between the exposure to uranium and a specific health outcome in humans. The evidence fulfills the criteria for sufficient evidence of an association (below) and satisfies several of the criteria used to assess causality: strength of association, dose-response relationship, consistency of association, temporal relationship, specificity of association, and biological plausibility.

Sufficient Evidence of an Association

Evidence is sufficient to conclude that there is an association. That is, a consistent association unlikely to be due to sampling variability has been observed between exposure to uranium and a specific health outcome in human studies that were free of severe bias and that controlled for confounding.

Limited/Suggestive Evidence of an Association

Evidence is suggestive of an association between exposure to uranium and a specific health outcome, but the body of evidence is limited by insufficient avoidance of bias, insufficient control for confounding, or large sampling variability.

Inadequate/Insufficient Evidence to Determine Whether an Association Exists

Evidence is of insufficient quantity, quality, or consistency to permit a conclusion regarding the existence of an association between exposure to uranium and a specific health outcome in humans.

Limited/Suggestive Evidence of No Association

Evidence is consistent in not showing an association between exposure to uranium of any magnitude and a specific health outcome. A conclusion of no association is inevitably limited to the conditions, magnitudes of exposure, and length of observation in the available studies.

REFERENCES

Alpert, J. S., and R. J. Goldberg. 2007. Dear patient: Association is not synonymous with causality. *American Journal of Medicine* 120(8):649-650.

Ellwood, J. M. 1998. *Critical appraisal of epidemiological studies and clinical trials*. 2nd ed. Oxford, UK: Oxford University Press.

Evans, A. S. 1976. Causation and disease: The Henle-Koch postulates revisited. *Yale Journal of Biology and Medicine* 49(2):175-195.

Hill, A. B. 1965. The environment and disease: Association or causation? *Proceedings of the Royal Society of Medicine* 58:295-300.

IARC (International Agency for Research on Cancer). 1998. *Preamble to the IARC monographs*. http://www-cie.iarc.fr/monoeval/background.html (accessed July 19, 2005).

———. 2004. *Tobacco smoke and involuntary smoking. IARC monographs on the evaluation of carcinogenic risks to humans*. Lyon, France: International Agency for Research on Cancer.

———. 2005. *Monographs on the evaluation of carcinogenic risks to humans*. http://www-cie.iarc.fr/ (accessed August 3, 2005).

———. 2006a. *Evaluation and rationale. Preamble to the IARC monographs*. http://monographs.iarc.fr/ENG/Preamble/currentb6evalrationale0706.php (accessed June 10, 2008).

———. 2006b. *Objective and scope. Preamble to the IARC monographs.* http://monographs.iarc.fr/ENG/Preamble/currenta2objective0706.php (accessed June 10, 2008).

IOM (Institute of Medicine). 1991. *Adverse effects of pertussis and rubella vaccines.* Washington, DC: National Academy Press.

———. 1994a. *Adverse events associated with childhood vaccines: Evidence bearing on causality.* Washington, DC: National Academy Press.

———. 1994b. *Veterans and Agent Orange: Health effects of herbicides used in Vietnam.* Washington, DC: National Academy Press.

———. 1996. *Veterans and Agent Orange: Update 1996.* Washington, DC: National Academy Press.

———. 1999. *Veterans and Agent Orange: Update 1998.* Washington, DC: National Academy Press.

———. 2000a. *Clearing the air: Asthma and indoor air exposures.* Washington, DC: National Academy Press.

———. 2000b. *Gulf War and health, volume 1: Depleted uranium, sarin, pyridostigmine bromide, vaccines.* Washington, DC: National Academy Press.

———. 2001. *Veterans and Agent Orange: Update 2000.* Washington, DC: National Academy Press.

———. 2003. *Veterans and Agent Orange: Update 2002.* Washington, DC: The National Academies Press.

———. 2004. *Gulf War and health: Updated literature review of sarin.* Washington, DC: The National Academies Press.

———. 2005. *Gulf War and health, volume 3: Fuels, combustion products, and propellants.* Washington, DC: The National Academies Press.

———. 2006. *Gulf War and health, volume 4: Health effects of serving in the Gulf War.* Washington, DC: The National Academies Press.

———. 2007. *Gulf War and health, volume 5: Infectious diseases.* Washington, DC: The National Academies Press.

Monson, R. 1990. *Occupational epidemiology.* 2nd ed. Boca Raton, FL: CRC Press.

NRC (National Research Council). 2006. *Human biomonitoring for environmental chemicals.* Washington, DC: The National Academies Press.

Susser, M. 1973. *Casual thinking in the health sciences: Concepts and strategies of epidemiology.* New York: Oxford University Press.

———. 1977. Judgment and causal inference: Criteria in epidemiologic studies. *American Journal of Epidemiology* 105(1):1-15.

———. 1988. Falsification, verification, and causal inference in epidemiology: Reconsideration in the light of Sir Karl Popper's philosophy. In *Causal inference*, edited by K. J. Rothman. Chestnut Hill, MA: Epidemiology Resources. Pp. 33-58.

———. 1991. What is a cause and how do we know one? A grammar for pragmatic epidemiology. *American Journal of Epidemiology* 133(7):635-648.

U.S. Surgeon General. 2004. *The health consequences of smoking: A report of the surgeon general.* http://www.surgeongeneral.gov/library/smokingconsequences (accessed October 26, 2004).

Wegman, D. H., N. F. Woods, and J. C. Bailar. 1997. Invited commentary: How would we know a Gulf War syndrome if we saw one? *American Journal of Epidemiology* 146(9):704-711.

5

Exposure Assessment

Human exposure assessment is a key component in understanding whether depleted uranium can cause specific health outcomes. Most of this chapter summarizes three reports—the "Capstone report" (USACHPPM, 2004), the "Sandia report" (Marshall, 2005), and the "Royal Society report" (Royal Society, 2001). Those reports used models to estimate depleted-uranium exposures in military personnel of the Gulf War. Information about exposures to depleted uranium that probably occurred in Gulf War troops will help to put health outcomes observed in epidemiologic studies described in Chapter 7 into context. There are several methods for monitoring depleted uranium in the body, and the reminder of this chapter describes how depleted uranium is detected in humans.

ESTIMATION OF EXPOSURE TO DEPLETED URANIUM DURING THE GULF WAR

Depleted-uranium penetrator strikes can produce inhalable aerosols that contain high concentrations of depleted uranium and depleted-uranium fragments that can cause shrapnel wounds (DOD, 2000). Substantial amounts of aerosol can be generated when a depleted-uranium penetrator strikes military vehicles, such as Abrams tanks and Bradley vehicles. The residual level of depleted uranium is affected by the vehicles' ventilation rates. Ingestion of depleted-uranium particles in contaminated vehicles is possible but probably is not an important exposure route.

Exposure to depleted-uranium aerosols can be affected by characteristics of the struck vehicle, impact conditions, and residence time of personnel in the contaminated vehicle. If the target is soft (such as a lightly armored vehicle), the depleted-uranium penetrator is likely to pass through it with little conversion of the penetrator rod into depleted-uranium oxides (Royal Society, 2001). If a penetrator strikes the depleted-uranium armor of a modern battle tank, much of the penetrator mass will be converted into depleted-uranium oxides. When a modern battle tank, such as an Abrams tank, is involved in a "friendly-fire" incident, substantial amounts of depleted-uranium aerosols can be generated by the impact.

During friendly-fire incidents, various exposure scenarios occurred. To categorize the exposure levels, exposures to depleted uranium during the Gulf War have been classified into three categories (DOD, 2000; USACHPPM, 2000), which provide a useful framework for considering potential depleted-uranium intakes and associated risks that was used in the Capstone report (USACHPPM, 2004), the Sandia report (Marshall, 2005), and the Royal Society report (Royal Society, 2001). The three categories are defined here.

- Level I includes military personnel in, on, or near combat vehicles at the time of impact and perforation by depleted-uranium munitions or personnel who entered vehicles immediately after they were struck (and perforated) by depleted-uranium munitions. The personnel could have been exposed to depleted uranium by contact with fragments resulting from impact or their being embedded in the body, by inhalation of depleted-uranium aerosols, by ingestion of depleted-uranium residues, or by settling of depleted uranium particles on open wounds, burns, or other breaks in the skin—or by any combination of these possibilities. This level also includes personnel occupying a vehicle whose depleted-uranium armor is perforated by non–depleted-uranium munitions.
- Level II includes military personnel and a small number of Department of Defense (DOD) civilian employees whose job functions required them to work in and around vehicles that contained depleted-uranium fragments and particles. Those people were not in a vehicle at the time of impact and did not immediately enter it after it was struck. They performed a variety of tasks, such as battle-damage assessment, repairs, explosive-ordnance disposal, and intelligence-gathering. They typically entered vehicles well after the initial suspended aerosol had dissipated or settled onto interior surfaces. They may have inhaled depleted-uranium residues that were resuspended by their physical activities, ingested depleted uranium through hand-to-mouth transfer, or spread contamination on their clothing. DOD personnel who were involved in cleaning up depleted-uranium residues generated during other events, such as the July 11, 1991, explosion and fires at the Camp Doha North Compound, are also included in this group.
- Level III is an "all others" group whose exposures were brief or incidental. This group includes personnel who entered depleted-uranium–contaminated Iraqi equipment, were downwind of burning Iraqi or US equipment struck by depleted-

uranium rounds, or were downwind of burning depleted-uranium ammunition (such as personnel at Camp Doha during the July 11, 1991, explosions and fire). Although these people could have inhaled airborne depleted-uranium particles, they are unlikely to have received an intake high enough to cause health effects.

Level III exposure is likely to be much lower than level I or II exposure. However, the number of military personnel with level III exposure may be much larger and the exposure range wider.

Direct measurement of exposure to depleted uranium in the battlefield is ideal but is not practical. To estimate exposure, field tests have been conducted to measure the range of depleted-uranium concentrations in vehicles that have been struck by large-caliber depleted-uranium rounds (USACHPPM, 2004). Aerosols were collected while Abrams tank and Bradley vehicle ballistic hulls and turrets with depleted-uranium armor or conventional armor were struck by depleted-uranium rounds. When an Abrams tank with depleted-uranium armor was struck, inhalation intake of depleted-uranium oxides by surviving crew members in 5 minutes is estimated to have been 20% greater than intake by the crew of a tank with conventional armor. When the less heavily armored Bradley vehicle was struck, inhalation intake was estimated to be 30% of the intake in the Abrams tank.

Residual concentration in a struck vehicle can be affected by ventilation of the vehicle. In one field test, inhalation intake of depleted uranium was reduced by about 90% in a struck Abrams tank with an operating ventilation system compared with that in a tank without one (USACHPPM, 2004). However, it is not clear whether the ventilation systems were active during friendly-fire incidents in the Gulf War. Without confirmed information on ventilation, estimation of exposure should be based on the cautious assumption of no ventilation.

All three reports mentioned above—the Capstone report, the Sandia report, and the Royal Society report—were based on an estimation approach, so they are all subject to considerable uncertainties in intake estimates and due to parameters chosen for modeling. The Royal Society report used the best data then available on initial air concentrations of depleted-uranium oxides in a struck tank and produced central estimates of intakes and risks for a number of exposure scenarios. Worst-case estimates were also provided by using values at the upper end of the likely range. The worst-case scenarios provide intakes and risks that are unlikely to be exceeded.

The Capstone report addressed inadequacy of the available data from test firings by conducting 13 new test firings of large-caliber depleted-uranium rounds against an Abrams tank and a Bradley vehicle. The Capstone study determined the airborne depleted-uranium concentration and size distribution in the struck vehicles as functions of time after impact. It provides a substantial database of airborne depleted-uranium concentration in struck vehicles and of the composition and particle size distribution. In addition, in vitro solubility of particles in the aerosol allowed estimates of the likely intake of depleted uranium and health risk. A National Research Council committee reviewed and evaluated the Capstone report and concluded that the "methods and results of the Capstone exposure assessment to be appropriate and well done" (NRC, 2008).

The Sandia report independently analyzed the Capstone test-firing data, estimated intakes by soldiers on the battlefield, and provided predicted risks to health. The Capstone and Sandia reports provide typical intakes and maximum intakes based on the highest observed value in test firings. Direct comparison of the three reports is complicated by use of different methods, terminologies, and parameters. However, all three reports agree on the general extent of the intakes from and health risks posed by depleted uranium.

Level I Exposures

Each of the three reports estimated level I exposure through inhalation by modeling with data from test firings. The data from the Capstone test firings showed a range of inhalation intakes of 250-710 mg of depleted uranium by the surviving crews of an Abrams tank struck by a single large-caliber depleted-uranium penetrator. The range for first responders was 150-200 mg. The Sandia study reported a best estimate of 250 mg and a maximum of 4 g on the basis of the Capstone test-firing data, which are almost identical with those in the Royal Society report.

The three reports had similar estimates of peak renal uranium concentration. A comparison of the level I exposure estimates is shown in Table 5-1. The best estimates of 3 μg/g of kidney and up to 6.5 μg/g of kidney were reported in the Sandia study and the Capstone study, respectively. The central peak of 4 μg/g of kidney was reported in the Royal Society report. The worst-case renal concentration of 400 μg/g in the Royal Society study and the maximum of 53 μg/g in the Sandia study suggest that kidney failure is possible under extreme circumstances.

It is generally accepted that the risk of developing cancer is related to the radiologic dose. Internal exposure to alpha radiation, such as from deposited depleted-uranium aerosols, may increase the risk of cancer. Each report provided an effective dose based on a period of 50 years after exposure.

The predicted increase in lifetime risk of death from lung cancer from level I exposure is about 0.1% in the Royal Society and Sandia reports and 0.06-0.3% in the Capstone report. An increase risk of 0.1% implies that the chance over a person's lifetime of dying of lung cancer from level I exposure is 0.1% greater than the background rate of cancer mortality. The worst-case estimate of lung cancer in the Royal Society report was 6.5%. The maximum estimates were 3.5% and 0.4-1.4% in the Sandia report and the Capstone report, respectively. The increase in lifetime risk of other fatal cancers, including leukemia, is much lower than the increase in risk of lung cancer.

TABLE 5-1 Comparison of Level I Exposure Estimates and Risk

	Capstone (Maximum)	Sandia (Maximum)	Royal Society (Worst Case)
Estimated inhalation intake of depleted uranium	160-710 mg (1 g)	250 mg (4 g)	250 mg (5 g)
Peak renal uranium	0.7-6.5 (16) µg/g of kidney	3 (53) µg/g of kidney	4 (400) µg/g of kidney
Effective dose	0.9-6.0 rems or 9-60 mSv	15 mSv (250) by inhalation only 19 mSv (265) by inhalation and fragment	22 mSv (1,100) Inhalation and ingestion
Excess lifetime risk of lung cancer[a]	0.06-0.3% (0.4%)[c]	0.085% (1.4%)	0.12% (6.5%)
Excess lifetime risk of leukemia[b]	NA	0.0004% (0.007%)	0.0005% (0.005%)

NOTE: NA = not available.
[a]Lifetime risk of lung cancer: 7.91% in US men, 6.18% in US women (ACS, 2008).
[b]Lifetime risk of leukemia: 1.50% in US men, 1.06% in US women (ACS, 2008).
[c]Spratt, 2007.

In addition to inhalation intake, depleted-uranium shrapnel wounds constitute a potential exposure route for those involved in level I exposure scenarios. In the Gulf War, 6 Abrams tanks and 15 Bradley vehicles were involved in friendly-fire incidents. The total number of soldiers surviving those incidents was 104. The Baltimore Department of Veterans Affairs health-surveillance program recruited and followed this depleted-uranium–exposed cohort. A total of 74 soldiers have participated in at least one visit since 1993. Of the 74, 19 have evidence of retained shrapnel as indicated by skeletal X-ray analysis.

Whole-body radiation counting was conducted in 29 depleted-uranium–exposed soldiers, including those with shrapnel (McDiarmid et al., 2000). Only nine have detectable scores above the background provided by the counting chamber, and all nine had shrapnel. The lack of sensitivity may be due to the low radioactivity of uranium and the tissue absorption of depleted-uranium radiation.

Another way to estimate uranium exposure is on the basis of urinary uranium excretion. Urinary uranium concentration can be a biomarker of total cumulative dose. An occupational-exposure decision level of 0.8 µg/L is used by the Department of Energy Fernald Environmental Management Project (FEMP, 1997). The soldiers with shrapnel continue to excrete high levels of uranium. The mean concentration in depleted-uranium–exposed soldiers is well above an upper-limit value that occurs in a normal population owing to intake of natural uranium in drinking water (0.365 μg/L) (ICRP, 1974) and above the occupational-exposure decision level of 0.8 μg/L as a trigger for investigating work areas for unsuspected high exposure to uranium.

Radiation dose to the depleted-uranium–exposed cohort was estimated by using urinary uranium-excretion data for a 10-year period after the Gulf War (Squibb et al., 2005). The upper bound of estimated lifetime (50-year) radiation dose of the depleted-uranium–exposed soldier who had the highest urinary concentration was 60 mSv, which is close to the National Council on Radiation Protection and Measurements allowable radiation dose for the public of 50 mSv (NCRP, 1993) and the US Nuclear Regulatory Commission's regulations for occupational dose to individual adults of 50 mSv/year (10 CFR 20.1201).

Level II Exposures

Level II exposures depend on the amount of time spent working in contaminated vehicles. The exposure estimates are based on a single acute intake and assume that personal protective equipment is not used and that decontamination has not taken place before the person's entry into the vehicle. Central or best estimates are based on 10 hours of work in contaminated vehicles. Estimated inhalation intakes in the Capstone, Sandia, and Royal Society reports were 5-40 mg, as shown in Table 5-2. Peak renal uranium concentrations were 0.03-0.5 μg/g of kidney. The Capstone study reported effective dose in rems per hour; this requires that the total time in the vicinity of the depleted-uranium–perforated vehicle and the fraction of time in the vehicle be known. The Sandia and Royal Society studies estimated effective doses from depleted-uranium exposure scenarios based on a 50-year postexposure duration. Lung-cancer risks are at least fivefold less than the corresponding level I estimate. Maximum and worst-case level II estimates are based on 100 hours of work in a contaminated vehicle. Excess lifetime lung-cancer risks are 0.2-0.4% in the Capstone and Sandia reports and 2.4% in the Royal Society report.

Level III Exposures

Level III exposures can result from briefly entering a contaminated vehicle, from exposure to plumes downwind of penetrator impacts, and from exposure to resuspended soil. The health risks associated with level III exposure are predicted to be very low. Estimated inhalation intakes in the Capstone, Sandia, and Royal Society reports were 0.5-6 mg, as shown in Table 5-3. Most soldiers on the battlefield may have level III exposure or less. Peak renal uranium concentration in a worst-case scenario could lead to some renal dysfunction. The excess risks

TABLE 5-2 Comparison of Level II Exposure Estimates and Risk

	Capstone (Maximum)	Sandia (Maximum)	Royal Society (Worst Case)
Estimated inhalation intake of depleted uranium	5 mg (1.45 g)	40 mg (0.6 g)	10 mg (2 g)
Peak renal uranium concentration	0.03 (14) μg/g of kidney	0.5 (8) μg/g of kidney	0.05 (96) μg/g of kidney
Effective dose	0.00197 rem/hour (0.078) Additional 0.000707 rem/hour by ingestion	2.5 mSv (38)	0.52 mSv (440)
Excess lifetime risk of lung cancer[a]	0.001% (0.4%)	0.014% (0.21%)	0.0025% (2.4%)
Excess lifetime risk of leukemia[b]	NA	0.0001% (0.001%)	<0.00001% (0.001%)

NOTE: NA = not available.
[a]Lifetime risk of lung cancer: 7.91% in US men, 6.18% in US women (ACS, 2008).
[b]Lifetime risk of leukemia: 1.50% in US men, 1.06% in US women (ACS, 2008).

TABLE 5-3 Comparison of Level III Exposure Estimates and Risk

	Capstone (Maximum)	Sandia (Maximum)	Royal Society (Worst Case)
Estimated inhalation intake of depleted uranium	0.5 mg (145 mg)	6 mg (60 mg)	1 mg (200 mg)
Peak renal uranium	0.003 (1.4) μg/g of kidney	Negligible (0.08) μg/g of kidney	0.005 (10) μg/g of kidney
Effective dose	0.00197 rem/hour (0.078)	10-5 mSv (0.05)	0.09 mSv (66)
	Additional 0.00012 rem/hour by ingestion		From all pathways
Excess lifetime risk of lung cancer[a]	0.0001% (0.04%)	<0.0001% (0.002%)	<0.0001% (0.02%)
Excess lifetime risk of leukemia[b]	NA	<0.00001% (0.00001%)	<0.00001% (0.00001%)

NOTE: NA = not available.
[a]Lifetime risk of lung cancer: 7.91% in US men, 6.18% in US women (ACS, 2008).
[b]Lifetime risk of leukemia: 1.50% in US men, 1.06% in US women (ACS, 2008).

of lung cancer and leukemia from level III exposure were less than 0.0001% and less than 0.00001%, respectively.

In addition to battlefield exposure, there are concerns about long-term exposure of residents in areas where depleted-uranium munitions were deployed. Depleted-uranium penetrators might miss their intended targets and end up embedded several feet in the ground, so they could lead to increased uranium concentrations in soil and water supplies. Some depleted-uranium oxides can be resuspended and cause inhalation exposure. Such environmental exposure can cause long-term exposure in the local population. Estimation of such exposure requires understanding of different exposure pathways and information about environmental contamination.

The Royal Society and Sandia reports estimate intakes and risks in the general population where depleted-uranium munitions were deployed. Cancer risks to the local population from long-term inhalation are estimated to be extremely low in both reports, even in the worst-case scenario. A worst-case estimate of uranium at 0.1-0.2 μg/g of kidney was reported by the Royal Society. High concentrations of depleted uranium around sites of penetrator impacts may present some risks to children playing in these areas for long periods. The Sandia report estimates an excess lifetime risk of fatal lung cancer of 0.035% in children who play for 300 hours in and 700 hours outside a depleted-uranium–contaminated vehicle.

EXPOSURE-MONITORING METHODS

Occupational exposure and environmental exposure to depleted uranium as used in the military are the primary subjects of this report. Military personnel may be at risk of inhaling airborne depleted-uranium particles, ingesting depleted-uranium particles from contaminated vehicles, and having wounds become contaminated with depleted-uranium particles. As summarized in Chapter 3, toxicologic studies have demonstrated that some health effects can occur in uranium-exposed animal models. Therefore, the committee's primary interest in uranium exposure is based on its action as an internal toxicant.

Characterization of exposure to depleted uranium should include both radiologic and chemical exposures, since the radiologic and chemical properties of depleted uranium could act synergistically to cause adverse health outcomes. External radiation exposure can be measured by a personal film badge, which can record exposure due to gamma rays, X-rays, and beta particles. Thermoluminescent dosimeters are also used to record the cumulative exposure of workers over a predetermined period. External chemical exposure to uranium can be measured on the basis of airborne concentration. Such measurements can use stationary monitoring in workplaces or personal exposure monitoring. However, because the chemical toxicity of a uranium compound can be affected by its solubility, airborne uranium concentration is not commonly used in epidemiologic studies.

The internal dose resulting from exposure to uranium can be measured with biologic monitoring. Several methods are available for measuring uranium in biologic specimens (fluids, such as urine, and tissues, such as blood, hair, and nails). The methods include thermal ionization mass spectrometry (TIMS), instrumental neutron-activation analysis, delayed neutron counting, inductively coupled plasma-mass spectrometry (ICP-MS), inductively coupled plasma atomic-emission spectroscopy, α-spectroscopy, spectrophotometry, fluorometry, and kinetic phosphorescence analysis. Todorov et al. (2007) reviewed and compared those methods. Because of its accuracy, precision, high sample throughput, and ease of use, ICP-MS has become the preferred method for measuring uranium and depleted uranium in biologic samples (for example, Berard et al., 2003; Roth et al., 2003; Westphal et al., 2004; Ejnik et al., 2005; Parrish et al., 2006; Todorov et al., 2007). However, Horan et al. (2002) reported that TIMS has the "lowest detection limits of all current methods" and "is in the category of the best analytical method for uranium isotope determination in biological specimens." Assessment of internal dose of uranium compounds is usually based on urinalysis (Ejnik et al., 2005), but hair and nail analysis can also be used (Karpas, 2001; Karpas et al., 2005).

Whole-body radioactivity counting can detect small amounts of radioactive material (McDiarmid et al., 2000). However, it was not sensitive enough to detect depleted-uranium body burden in some depleted-uranium–exposed soldiers (Toohey, 2003). Only 9 of 29 depleted-uranium–exposed soldiers had detectable scores above the background level. The insensitivity of the method is due to the low radioactivity of depleted uranium and the tissue absorption of depleted-uranium radiation that was measured with a tissue-equivalent phantom containing known amounts of depleted uranium at different depths.

Uranium is widely present in the natural environment. The general population can be exposed to a natural background level of uranium. The third *National Report on Human Exposure to Environmental Chemicals* (CDC, 2005) reported a geometric mean urinary uranium excretion of 9 ng/L in a sample of about 5,000 people across the United States; 95% of the population had concentrations below 46 ng/L. Studies of nonoccupationally exposed persons have shown uranium concentrations in the general population of 11-22 ng/L (Dang et al., 1992; Medley et al., 1994; Ting et al., 1999). Urinary uranium concentrations of 1-41,800 ng/g of creatinine have been measured in soldiers and veterans where depleted uranium has been used (McDiarmid et al., 2006). A concentration of 1 ng/g of creatinine is equal to 1 ng/L of urine (Melissa McDiarmid, personal communication, June 28, 2007).

Internal dose can be reconstructed from bioassay data by using a biokinetic model that predicts the time-dependent distribution and excretion of radionuclides deposited in the human body. A generic respiratory tract model can describe the deposition and retention of inhaled material in the respiratory tract and its clearance to blood or to the gastrointestinal tract. The Human Respiratory Tract Model developed by the International Commission on Radiological Protection can predict the behavior of inhaled radionuclides in the respiratory tract (ICRP, 2002). In the case of uranium, one of three generic absorption types can be applied for the chemical and physical form of the inhaled element. That approach may be unreliable because the assumption of different absorption rates can cause errors in the estimation of internal dose. The upper-bound estimated lifetime (50-year) radiation dose to the depleted-uranium–exposed soldier with the highest urinary uranium concentration was 0.06 Sv (Squibb et al., 2005).

REFERENCES

ACS (American Cancer Society). 2008. *Cancer facts and figures.* Atlanta, GA: American Cancer Society.

Berard, P., A. Montegue, F. Briot, and J. C. Harduin. 2003. Procorad's international intercomparisons highlight the evolution of techniques used to determine uranium in urine. *Radiation Protection Dosimetry* 105(1-4):447-450.

CDC (Centers for Disease Control and Prevention). 2005. *Third national report on human exposure to environmental chemicals.* Atlanta, GA: Department of Health and Human Services, Centers for Disease Control and Prevention.

Dang, H. S., V. R. Pullat, and K. C. Pillai. 1992. Determining the normal concentration of uranium in urine and application of the data to its biokinetics. *Health Physics* 62(6):562-566.

DOD (Department of Defense). 2000. *Environmental exposure report: Depleted uranium in the Gulf (II).* Washington, DC: Department of Defense.

Ejnik, J. W., T. I. Todorov, F. G. Mullick, K. Squibb, M. A. McDiarmid, and J. A. Centeno. 2005. Uranium analysis in urine by inductively coupled plasma dynamic reaction cell mass spectrometry. *Analytical and Bioanalytical Chemistry* 382(1):73-79.

FEMP (Federal Environmental Management Project). 1997. *Technical basis for internal dosimetry at the Federal Environmental Management Project.* Fernald, OH: FEMP Publication.

Horan, P., L. Dietz, and A. Durakovic. 2002. The quantitative analysis of depleted uranium isotopes in British, Canadian, and U.S. Gulf War veterans. *Military Medicine* 167(8):620-627.

ICRP (International Commission on Radiological Protection). 1974. *Report of the task groups on reference man.* Elmsford, NY: Pergamon Press.

———. 2002. *Supporting guidance 3. Guide for the practical application of the ICRP human respiratory tract model.* Oxford, UK: Pergamon Press.

Karpas, Z. 2001. Uranium bioassay—beyond urinalysis. *Health Physics* 81(4):460-463.

Karpas, Z., A. Lorber, H. Sela, O. Paz-Tal, Y. Hagag, P. Kurttio, and L. Salonen. 2005. Measurement of the 234U/238U ratio by MC-ICPMS in drinking water, hair, nails, and urine as an indicator of uranium exposure source. *Health Physics* 89(4):315-321.

Marshall, A. C. 2005. Analysis of uranium dispersal and health effects using a Gulf war case study. *Government Reports Announcements and Index* (23):212.

McDiarmid, M. A., J. P. Keogh, F. J. Hooper, K. McPhaul, K. Squibb, R. Kane, R. DiPino, M. Kabat, B. Kaup, L. Anderson, D. Hoover, L. Brown, M. Hamilton, D. Jacobson-Kram, B. Burrows, and M. Walsh. 2000. Health effects of depleted uranium on exposed Gulf War veterans. *Environmental Research* 82(2):168-180.

McDiarmid, M. A., S. M. Engelhardt, M. Oliver, P. Gucer, P. D. Wilson, R. Kane, M. Kabat, B. Kaup, L. Anderson, D. Hoover, L. Brown, R. J. Albertini, R. Gudi, D. Jacobson-Kram, C. D. Thorne, and K. S. Squibb. 2006. Biological monitoring and surveillance results of Gulf War I veterans exposed to depleted uranium. *International Archives of Occupational and Environmental Health* 79(1):11-21.

Medley, D. W., R. L. Kathren, and A. G. Miller. 1994. Diurnal urinary volume and uranium output in uranium workers and unexposed controls. *Health Physics* 67(2):122-130.

NCRP (National Council on Radiation Protection and Measurements). 1993. *Limitation of exposure to ionizing radiation. Report number 116*. Bethesda, MD: National Council on Radiation Protection and Measurements.

NRC (National Research Council). 2008. *Review of toxicologic and radiologic risks to military personnel from exposure to depleted uranium during and after combat*. Washington, DC: The National Academies Press.

Parrish, R. R., M. F. Thirlwall, C. Pickford, M. Horstwood, A. Gerdes, J. Anderson, and D. Coggon. 2006. Determination of 238U/235U, 236U/238U and uranium concentration in urine using SF-ICP-MS and MC-ICP-MS: An interlaboratory comparison. *Health Physics* 90(2):127-138.

Roth, P., V. Hollriegl, E. Werner, and P. Schramel. 2003. Assessment of exposure to depleted uranium. *Radiation Protection Dosimetry* 105(1-4):157-161.

Spratt, B. 2007. Health hazards of depleted uranium munitions: Estimates of exposures and risks in the Gulf War, the Balkans, and Iraq. In *Depleted uranium: Properties, uses, and health consequences*, edited by A. C. Miller. Boca Raton, FL: CRC Press. Pp. 121-141.

Squibb, K. S., R. W. Leggett, and M. A. McDiarmid. 2005. Prediction of renal concentrations of depleted uranium and radiation dose in Gulf War veterans with embedded shrapnel. *Health Physics* 89(3):267-273.

Royal Society. 2001. *The health hazards of depleted uranium munitions: Part I*. London, UK: The Royal Society Working Group on the Health Hazards of Depleted Uranium Munitions.

Ting, B. G., D. C. Paschal, J. M. Jarrett, J. L. Pirkle, R. J. Jackson, E. J. Sampson, D. T. Miller, and S. P. Caudill. 1999. Uranium and thorium in urine of United States residents: Reference range concentrations. *Environmental Research* 81(1):45-51.

Todorov, T. I., J. W. Ejnik, F. G. Mullick, and J. A. Centeno. 2007. Chemical and histological assessment of depleted uranium in tissues and biological fluids. In *Depleted uranium: Properties, uses, and health consequences*, edited by A. C. Miller. Boca Raton, FL: CRC Press. Pp. 85-103.

Toohey, R. E. 2003. Excretion of depleted uranium by Gulf War veterans. *Radiation Protection Dosimetry* 105(1-4):171-174.

USACHPPM (U.S. Army Center for Health Promotion and Preventive Medicine). 2000. *Health risk assessment consultation: Depleted uranium-human exposure assessment and health risk characterization in support of the environmental exposure report. "Depleted uranium in the Gulf" of the office of the special assistant to the secretary of defense for Gulf War illnesses, medical readiness and military deployments. Osagwi levels I, II and III scenarios.* http://www.gulflink.osd.mil/chppm_du_rpt_index.html (accessed November 2, 2007).

———. 2004. *Capstone report: Depleted uranium aerosol doses and risks: Summary of U.S. Assessments*. Fort Belvoir, VA: U.S. Army Heavy Metals Office Chemical and Biological Defense Information Analysis Center.

Westphal, C. S., J. A. McLean, S. J. Hakspiel, W. E. Jackson, D. E. McClain, and A. Montaser. 2004. Determination of depleted uranium in urine via isotope ratio measurements using large-bore direct injection high efficiency nebulizer-inductively coupled plasma mass spectrometry. *Applied Spectroscopy* 58(9):1044-1050.

6

Clinical End Points of Interest

This chapter identifies the health outcomes on the basis of which the committee draws conclusions about the long-term human health effects associated with exposure to natural and depleted uranium (the committee's conclusions are presented in Chapter 8). It also provides background information, including incidence and prevalence rates and known risk factors, on each health outcome.

CANCER OUTCOMES

Cancer is a group of diseases characterized by uncontrolled growth and spread of abnormal cells (WHO, 2003; ACS, 2007b). Cancer can affect almost any tissue. It is caused by external factors (such as tobacco, chemicals, radiation, and infectious agents) and internal factors (such as inherited mutations, hormones, immune conditions, and mutations that result from metabolism). Such factors may act together or in combination to initiate or promote carcinogenesis (ACS, 2007b). In adults, a latent period of 10 years or more may elapse between exposure or mutation and the detection of cancer.

Worldwide, among the member states of the World Health Organization, cancer is the third-leading cause of death after heart disease and infectious and parasitic diseases (WHO, 2003). Each year, cancer leads to 12% of deaths, or about 6 million people (WHO, 2003). In men, lung and stomach cancers are the most common worldwide; in women, breast, cervical, and lung cancers are the most common (WHO, 2003). In the United States, cancer is the second-leading cause of death, exceeded only by heart disease. An estimated 565,650 Americans are expected to die from cancer in 2008, accounting for about one-fourth of all deaths (ACS, 2007b). Lung cancer remains the leading cause of cancer deaths in both men and women; prostatic cancer is the most frequently diagnosed cancer in men, and breast cancer in women (ACS, 2007b).

Although several of the studies reviewed cancer at all sites, the committee chose to focus on selected sites, specifically leukemias, lymphomas, and lung, bone, renal, bladder, stomach, brain and other parts of the central nervous system, prostatic, and testicular cancer. Most of those may be found on the basis of the route of exposure (generally inhalation or ingestion) and the mechanism of clearance of radiologic and chemical toxicants. Testicular cancer, being the most common cancer among young men, is of special interest to Gulf War veterans, and some studies of veterans suggested a higher but nonsignificantly increased risk in them than in their nondeployed counterparts (IOM, 2006b). Prostatic cancer is the most frequently diagnosed cancer in men in the United States, and any slight increase in risk could result in large numbers of cases or deaths.

Lung Cancer (ICD-10[1] C34)

Lung cancer is the leading cause of cancer death in the United States and the second-most common cancer in both American men and women. An estimated 213,380 new cases of and 160,390 deaths from lung cancer were expected in 2007 in the United States, accounting for about 14.8% of all cancer diagnoses and 28.7% of all cancer deaths. Lung-cancer incidence in men has been declining substantially from a peak of 102 cases per 100,000 men in 1984; in women, the incidence is reaching a plateau after decades of increase (ACS, 2007b). In 2004 (the most recent year with available published incidence data), there were 60.0 new cases of lung cancer per 100,000 people in the United States (73.6 in men and 50.2 in women) and 53.3 deaths per 100,000 (70.3 in men and 40.9 in women) (Ries et al., 2007).

Lung cancer is classified into two main types based on the appearance of its cells. Non–small-cell lung cancer accounts for about 87% of all lung cancers and is divided into three subtypes based on size, shape, and chemical makeup: squamous-cell carcinoma (25-30% of all lung cancers) is linked to smoking and commonly found near a bronchus, adenocarcinoma (40% of all lung cancers) appears in the outer regions of the lungs, and large-cell undifferentiated carcinoma (10-15% of all lung cancers) is found in all areas and tends to metastasize quickly. The second main type, small-cell lung cancer, also known as oat-cell carcinoma, accounts for the remaining 10-15% and is almost always linked to smoking. Small-cell lung cancer originates primarily in the bronchi and tends to metastasize quickly throughout the body fairly early in the disease process. Tobacco-smoking is the predominant risk factor and is thought to account for about 87% of lung-cancer deaths. Other risk factors include exposures to such

[1]ICD-10 is the 10th edition of the *International Classification of Diseases*.

carcinogens as radon, asbestos, beryllium, silica, arsenic, and secondhand smoke; family history; and diet (ACS, 2007g).

Leukemias (ICD-10 C91-95)

Leukemias are malignant diseases that arise from precursor cells of white blood cells. An estimated 44,240 new cases of and 21,790 deaths from leukemia were expected in 2007 in the United States, accounting for about 3.1% of all cancer diagnoses and 3.9% of all cancer deaths (ACS, 2007b). In 2004, there were 12.0 new cases of leukemia per 100,000 people in the United States (15.4 in men and 9.5 in women) and 7.2 deaths per 100,000 (9.7 in men and 5.5 in women) (Ries et al., 2007).

Although all leukemias originate in the bone marrow, there are four main types, classified by the type and developmental stage of the cells involved. Leukemias can be acute—in which case the cells grow rapidly and are not able to mature—or chronic—in which case the cells grow and accumulate slowly and look mature. And leukemias can affect different types of cells: lymphocytic leukemias affect the lymphocytes, white blood cells that make up lymphoid tissue, and myeloid leukemias affect granulocytes or monocytes, white blood cells that circulate and protect the body against infection (ACS, 2007k). Acute lymphocytic leukemia affects children more frequently than adults, whereas chronic lymphocytic leukemia affects only adults, mostly over the age of 40 years (ACS, 2007d,f). Acute myeloid leukemia, also called acute nonlymphocytic leukemia, is the most common leukemia and usually affects adults, particularly men, although it can occur in children (ACS, 2007e). Chronic myeloid leukemia affects mostly adults and is rare in children (ACS, 2007a). The four types of leukemias can be divided into subtypes based on progression and cell subtypes.

Characterizing leukemia cases gathered retrospectively for epidemiologic studies and integrating the results of studies conducted over several decades are particularly challenging because successive diagnostic criteria, with corresponding groupings and nomenclature, have been used. Individual leukemias may have unique etiologic factors (for example, T-cell leukemia is caused by the retrovirus HTLV-I), but the recognized risk factors for leukemias in general include exposure to ionizing radiation or some chemicals (such as occupational exposure to benzene or chemotherapy with alkylating agents), some genetic conditions (such as some chromosomal abnormalities, including Down syndrome), and particular acquired blood diseases (for example, myelodysplastic syndromes may develop into acute myeloid leukemia) (NCI, 2003).

Lymphomas (ICD-10 C81-85)

This section discusses two types of lymphomas: Hodgkin's disease (HD; also called Hodgkin lymphoma) (ICD-10 C81) and non-Hodgkin lymphoma (NHL)

(ICD-10 C82-85). The lymph nodes are sites of particular concern because uranium is known to accumulate in them.

HD is a very rare cancer that originates in lymphatic tissue (ACS, 2007i). An estimated 8,190 new cases of and 1,070 deaths from HD were expected in 2007 in the United States, accounting for about 0.6% of all cancer diagnoses and 0.2% of all cancer deaths (ACS, 2007b). In 2004, there were 2.9 new cases of HD per 100,000 people in the United States (2.9 in men and 2.8 in women) and 0.4 death per 100,000 (0.5 in men and 0.3 in women) (Ries et al., 2007).

HD is a B-cell lymphoma characterized by microscopically identifiable Reed-Sternberg cells; all other cancers of the lymphatic tissues are NHL. The only known risk factors for HD are infectious mononucleosis (caused by the Epstein-Barr virus) and low immunity. HD has not been associated with family history, diet, or environmental exposure, including exposure to uranium.

An estimated 63,190 new cases of and 18,660 deaths from NHL were expected in 2007 in the United States, accounting for about 4.4% of all cancer diagnoses and 3.3% of all cancer deaths (ACS, 2007b). In 2004, there were 20.4 new cases of NHL per 100,000 people in the United States (24.7 in men and 17.1 in women) and 7.0 deaths per 100,000 (8.8 in men and 5.7 in women) (Ries et al., 2007).

NHL originates in the B cells or, less frequently, the T cells of the lymphatic tissue (ACS, 2007i). It encompasses the many types of lymphoma that remain after the exclusion of HD (B-cell lymphoma). In the evolving classification systems for lymphohematopoietic cancers overall, there have been a series of systems just for NHL. In ICD-10 coding, NHL is divided among ICD-10 C82-85. Characterizing NHL cases gathered retrospectively for epidemiologic studies and integrating the results of studies conducted over several decades are particularly challenging because of the nonconstancy in the terminology and coding for reporting diagnoses of or deaths from this family of diseases. Many risk factors have been identified for NHL: genetic or acquired severely compromised immune system; infection by HIV, related T-cell viruses, or Epstein-Barr virus or by some bacteria (such as *Helicobacter pylori* in the case of gastric lymphoma); aging; obesity, the only recognized "lifestyle" factor; radiation; chemotherapy drugs; and possibly some chemicals, with benzene, herbicides, and insecticides most often implicated.

Bone Cancer (ICD-10 C40-41)

An estimated 2,370 new cases of and 1,330 deaths from bone and joint cancer were expected in 2007 in the United States, accounting for about 0.2% of all cancer diagnoses and all cancer deaths (ACS, 2007b). During 2000-2004, there was 0.9 new case of bone and joint cancer per 100,000 people in the United States (1.0 in men and 0.8 in women) and 0.4 death per 100,000 (0.5 in men and 0.3 in women) (Ries et al., 2007).

Of the several forms of primary bone and joint cancer, osteosarcoma is the most common primary bone cancer, accounting for about 35% of all cases. Occurring more frequently in males, osteosarcoma is found mostly in people 10-30 years old and rarely during middle age. About 10% of cases develop in people 60 years old and older. Other rare forms of primary bone cancer include chondrosarcoma (cancer of cartilage cells), Ewing tumor (cancer of the bone cavity), chordoma (cancer of the skull base and spinal bones), and malignant fibrous histiocytoma and fibrosarcoma (cancer of the connective tissues). The 5-year survival rate can be as high as 80%, but the prognosis for people with primary bone cancer varies greatly, depending on the specific type of cancer and the stage at which it is diagnosed (NCI, 2002a; ACS, 2006b).

Risk factors for bone cancer are exposure to ionizing radiation, particularly at an early age or at high doses; a history of bone disorders, such as Paget disease; and the presence of multiple exostoses (overgrowths of bone tissue), multiple osteochondromas (benign bone tumors formed by bone and cartilage), multiple enchondromas (benign cartilage tumors), and some genetic factors (such as mutation of the p53 tumor-suppressor gene) (NCI, 2002a; ACS, 2006b).

Renal Cancer (ICD-10 C64-66)

An estimated 51,190 new cases of and 12,890 deaths from cancer of the kidney and renal pelvis were expected in 2007 in the United States, accounting for about 3.5% of all cancer diagnoses and 2.3% of all cancer deaths (ACS, 2007b). In 2004, there were 13.1 new cases of renal and renal pelvis cancer per 100,000 people in the United States (17.8 in men and 9.2 in women) and 4.1 deaths per 100,000 (5.9 in men and 2.7 in women) (Ries et al., 2007).

Over 90% of renal cancers in adults are renal-cell carcinomas (RCCs) or adenocarcinomas (ACS, 2007j). Most other malignant renal tumors are transitional-cell carcinomas that arise in the renal pelvis, ureter, or urethra; these are jointly referred to as urothelial carcinomas or cancers of the renal pelvis.

Smoking and obesity are the major risk factors for renal cancer. Others are diet, increasing age, male sex, some hereditary conditions (such as Von Hippel-Lindau disease and hereditary papillary renal-cell carcinoma), and dialysis treatment for renal disease. Such medications as phenacetin and diuretics (or the high blood pressure that they are used to treat) have also been associated with RCC, as has occupational exposure to asbestos, cadmium, and some organic solvents (ACS, 2007c).

Bladder Cancer (ICD-10 C67)

An estimated 67,160 new cases of and 13,750 deaths from urinary bladder cancer were expected in 2007 in the United States, accounting for about 4.6% of all cancer diagnoses and 2.5% of all cancer deaths (ACS, 2007b). In 2004, there were 20.6 new cases of urinary bladder cancer per 100,000 people in the United States (36.3 in men and 9.1 in women) and 4.4 deaths per 100,000 (7.6 in men and 2.2 in women) (Ries et al., 2007).

The bladder is lined with transitional and squamous cells. More than 90% of bladder cancers arise in transitional cells, and squamous-cell carcinomas make up only about 8% of bladder cancers (NCI, 2002b). Although cells that line the renal pelvis and ureter are histologically similar to bladder epithelial cells, tumors of the renal pelvis, ureters, and urethra are considered urothelial-cell tumors and are traditionally grouped with renal cancer. This section, however, addresses only bladder cancer.

The major risk factor for bladder cancer is smoking. Demographic factors that have some influence on the occurrence of bladder cancer are race (the incidence is highest in whites and lowest in Asians), increasing age, sex (males are at higher risk), and family history. Chronic bladder inflammation due to infections, bladder or kidney stones, and parasites has been associated with bladder cancer. Known occupational risk factors include use of the drug cyclophosphamide and exposure to aromatic amines, arsenic, and organic chemicals associated with manufacture of rubber, leather, textiles, and paint (ACS, 2006a).

Brain and Other Nervous System Cancers (ICD-10 C71-72)

An estimated 20,500 new cases of and 12,740 deaths from brain and other nervous system cancers were expected in 2007 in the United States, accounting for about 1.4% of all cancer diagnoses and 2.3% of all cancer deaths (ACS, 2007b). In 2004, there were 6.5 new cases of brain and other nervous system cancers per 100,000 people in the United States (7.6 in men and 5.5 in women) and 4.3 deaths per 100,000 (5.2 in men and 3.5 in women) (Ries et al., 2007).

Most nervous system tumors, including brain tumors, are not associated with known risk factors. The few known risk factors associated with these cancers are radiation, immune system disorders, and family history.

Stomach Cancer (ICD-10 C16)

Gastric cancer (commonly known as stomach cancer) was once a leading cause of cancer deaths in the United States; it is now more common in other countries. An estimated 21,260 new cases of and 11,210 deaths from stomach cancer were expected in 2007 in the United States, accounting for about 1.5% of all cancer diagnoses and 2.0% of all cancer deaths (ACS, 2007b). In 2004, there were 7.6 new cases of stomach cancer per 100,000 people in the United States (10.9 in men and 5.1 in women) and 4.0 deaths per 100,000 (5.5 in men and 2.8 in women) (Ries et al., 2007). *Helicobacter pylori* infection is a major cause of stomach cancer. Other risk factors include sex, age, ethnicity, diet, tobacco use, family history, and occupation (ACS, 2007m).

Male Genital Cancers (ICD-10 C61-62)

An estimated 218,890 new cases of and 27,050 deaths from prostatic cancer were expected in 2007 in the United States, accounting for about 28.5% of all cancer diagnoses and 9.3% of all cancer deaths in American men (ACS, 2007b). An estimated 7,920 new cases of and 380 deaths from cancer of the testis were expected, accounting for about 1% of all cancer diagnoses and 0.1% of all deaths in men. In 2004, there were 159.5 new cases of prostatic cancer per 100,000 men in the United States and 25.4 deaths per 100,000, and there were 5.7 new cases of testicular cancer per 100,000 and 0.2 death per 100,000 (Ries et al., 2007).

Factors that increase the risk of prostatic cancer include increasing age, race, and family history (ACS, 2007l). Testicular cancer is uncommon but highly treatable. Known or suspected risk factors include cryptorchidism, family history, some occupational exposures, multiple atypical nevi, HIV infection, race and ethnicity, body size, and maternal hormone use during pregnancy (ACS, 2007h).

NONCANCER OUTCOMES

Nonmalignant health outcomes selected by the committee for evaluation include renal disease, respiratory disease, neurologic effects, and reproductive effects. In addition to information on disease prevalence and risk factors, a discussion of clinical tests used to evaluate organ or organ-system function for renal and respiratory disease is included.

Nonmalignant Renal Disease (ICD-10 N18)

Chronic kidney disease (CKD), also known as chronic renal insufficiency, is the permanent loss of renal function (NIDDK, 2008). According to data from the National Health and Nutrition Examination Survey for 1994-2004, the prevalence of CKD in US adults 20 years old or older was 16.8% (NHANES, 2007). Diabetes and hypertension are leading risk factors for CKD, and many other factors have also been implicated, such as obesity, family history, ethnicity, race, and cardiovascular disease.

Stages of renal function are assigned according to the estimated glomerular filtration rate (GFR), a measure of the kidneys' capacity to filter toxins from the blood, and the presence of protein in the urine (proteinuria). Renal function is evaluated with numerous tests that broadly provide an indication of the GFR, the level of proteinuria, or how well the tubular portions of the kidneys' cellular structures are able to modify the fluid filtered from the blood as it is processed into urine (for example, on the basis of urinary pH, urinary glucose, and tubular protein markers). Abnormalities of the GFR and proteinuria are highly predictive not only of progressive kidney disease but of cardiovascular morbidity and mortality.

The GFR is most accurately measured with a clinical test in which an exogenous marker (for example, inulin, iohexol, or iothalamate) is injected into the body and then measured in carefully timed specimens of blood and urine as it is filtered and excreted by the kidneys. This type of testing is used principally in research settings because of the time and effort it demands. In the clinical setting, the attributes of the endogenous substance creatinine are often exploited to provide an indication of the GFR. Creatinine is a product of normal muscle metabolism and is principally excreted from the body by the kidneys through the process of filtration. Because creatinine production is typically constant (relative to a person's muscle mass) over short to medium periods, increases in blood creatinine concentration often indicate diminished GFR and therefore a loss of renal function. A more accurate way of using the creatinine concentration as a measure of renal function is to relate blood creatinine concentration to the quantity of creatinine excreted in the urine over a specified period (typically measured with a 24-hour urine collection). The quotient of urinary creatinine divided by plasma creatinine is known as the creatinine clearance. This value is a good proxy for the GFR measured with exogenous markers. Finally, to obviate the timed collection of urine, which can be inconvenient and difficult in the clinical setting, mathematical equations have been developed to estimate the GFR. The most commonly used equation is derived from the Modification of Diet in Renal Disease Study and incorporates age, race, sex, and serum creatinine to provide an estimate of GFR (eGFR). This metric appears to work reasonably well when the true GFR is 60 mL/min per 1.73 m^2 or lower, but it substantially underestimates the GFR in people who have more preserved renal function.

Normally, there is essentially no plasma protein in urine. The filtering structures in the kidneys (glomeruli) should permit only the aqueous portion of plasma and none of its protein to enter the urine. When the filtering structures become diseased, plasma protein can pass into the urine, a phenomenon that may or may not be accompanied by a decrease in the GFR. This form of glomerular proteinuria is predictive of poor clinical outcomes.

Measures of the function of the renal tubular structures are less commonly used in the clinical setting. Some, such as measures of tubular protein, have an uncertain relationship to clinical renal disease, and they do not clearly provide prognostic information with respect to the clinically important measures of GFR or glomerular proteinuria.

Finally, many other specialized tests can be used to assess specific aspects of the kidneys' ability to modify the filtered plasma as it is transformed into urine. They include measures of the electrolyte composition of the urine (for example, phosphate, calcium, sodium, and potassium), of the acidifying capabilities of the kidneys (for example, urinary pH), and of the kidneys' ability to appropriately reclaim small molecules, such as glucose, that are filtered but are not excreted in the urine of healthy people.

Nonmalignant Respiratory Disease (ICD-10 J43-46)

The category of nonmalignant respiratory disease includes several conditions. Chronic obstructive pulmonary disease (COPD), pneumonia, pneumoconiosis, and asthma are described here because they are common and an association between them and exposure to uranium is biologically plausible. COPD includes two frequently coexisting conditions: chronic bronchitis and emphysema. Chronic bronchitis is characterized by symptoms of cough and sputum production, which may be associated with poorly reversible pathologic changes in the airways. Emphysema is characterized by the progressive destruction of alveoli and permanent changes in the airways. COPD causes airflow obstruction that interferes with normal breathing. In 2005, an estimated 11.6 million adults in the United States had COPD; in 2004, 118,171 adults in the United States died from it (CDC, 2006). The primary risk factor for COPD is smoking; other risk factors include occupational exposure, sex, childhood history of respiratory infections, and family history (ALA, 2007a).

Pneumonia is an infection or inflammation of the lungs. It can affect a section or lobe of a lung (lobar pneumonia) or patches throughout both lungs (bronchopneumonia). Risk factors for pneumococcal pneumonia include chronic illness, recent recovery from severe illness, such specific environments as chronic-care facilities, and age. In 2004, deaths in the United States attributed to pneumonia numbered 58,564 (19.9 deaths per 100,000 people) (CDC, 2007c).

Pneumoconiosis is caused by exposure to inorganic dust. It may be functionally important, representing interstitial fibrosis, or have little or no functional impact.

Asthma is a chronic inflammatory condition in which acute exacerbations are caused by airway obstruction and inflammation; it is classified as a reversible obstructive lung disease caused by increased reaction of the airways to stimuli. Asthma interferes with breathing when the airways are narrowed because of swelling of the lining, tightening of the muscle, and increased mucus secretion. Its symptoms are coughing, wheezing, and shortness of breath, and they are triggered by such stimuli as respiratory infections, environmental pollutants, temperature change, and exercise (ALA, 2007b). In 2004, an estimated 15.7 million adults in the United States had asthma, and deaths in the United States attributed to asthma numbered 3,816 (1.3 deaths per 100,000 people) (CDC, 2007b).

Nonmalignant respiratory effects may be ascertained by several methods, including the following:

- Mortality: Deaths ascribed to such causes may be identified on the basis of death-certificate data.
- Physician diagnosis: Medical diagnosis of a nonmalignant respiratory disease may be determined from medical records or from patient reporting of the presence of a physician diagnosis.

• Symptoms: Symptoms consistent with a nonmalignant respiratory disorder may be identified by using standardized self-administered or interview-administered questionnaires.

• Pulmonary-function testing: Pulmonary-function testing, particularly spirometry, is commonly used to evaluate respiratory function clinically and in epidemiologic studies. Major measures include forced expiratory volume in 1 second (FEV1), forced vital capacity (FVC), and FEV1:FVC ratio. The latter is indicative of COPD and asthma. A statistical association between a pulmonary-function test result and a measured exposure suggests that the exposure agent has the potential to produce disease in people who are sufficiently exposed.

• Radiographic (imaging) studies: "Chest X-ray" testing is particularly useful for diagnosing interstitial lung diseases, such as pneumoconiosis. Such methods are widely used in studies of occupational cohorts exposed to agents other than uranium but have had only limited application to uranium-exposed workers.

Neurologic Effects (ICD-10 G00-G99) and Neurobehavioral and Neurocognitive Effects (ICD-10 F00-99)

There are numerous diseases of the nervous system, and, as described in Chapters 7 and 8, the epidemiologic studies of uranium-exposed populations do not specify individual neurologic diseases in their analyses. Prevalence, incidence, and risk factors vary among nervous system diseases. Several studies have suggested an increased risk of amyotrophic lateral sclerosis in Gulf War veterans (IOM, 2006a), but no risk factor has been identified. Neurocognitive and neurobehavioral outcomes have been assessed in a group of Gulf War veterans who were exposed to depleted uranium.

Birth Defects and Other Adverse Reproductive Outcomes (ICD-10 O00-Q99)

Birth defects occur in about one in 33 live births in the United States (CDC, 2007a). The numerous types of birth defects—several thousand have been identified—include structural defects, chromosomal abnormalities, and birth-defect syndromes (California Birth Defects Monitoring Program, 2006; March of Dimes, 2006). The most common birth defects in the United States are cleft palate, cleft lip, and Down syndrome (March of Dimes, 2006). Birth defects are caused by genetic and environmental factors (for example, chemicals and infectious agents) or a combination of such factors. The causes of most cases of birth defects are unknown (CDC, 2007a).

In addition to birth defects, adverse outcomes of pregnancy include early pregnancy loss (before 8 weeks), spontaneous abortion (8-20 weeks), ectopic pregnancy, and late fetal death and stillbirth. Such outcomes, especially early pregnancy loss and spontaneous abortion, are often not reported, so it is difficult to estimate their frequencies precisely. However, about 20-30% of implantations end in early pregnancy loss, and about 10-20% of clinically recognized pregnancies result in spontaneous abortion (NRC, 2000).

Adverse reproductive outcomes include abnormal male and female hormone profiles, altered menstrual and ovarian cycles, longer than normal time to pregnancy, abnormal semen characteristics, gynecologic and urologic disorders, and premature reproductive senescence (NRC, 2001). Information on the number of people affected by reproductive disorders is sparse.

REFERENCES

ACS (American Cancer Society). 2006a. *Detailed guide: Bladder cancer—what is bladder cancer?* http://www.cancer.org/docroot/CRI/CRI_2_3x.asp?dt=44 (accessed March 4, 2008).

———. 2006b. *Detailed guide: Bone cancer*. http://www.cancer.org/docroot/lrn/lrn_0.asp (accessed March 4, 2008).

———. 2007a. *All about leukemia—chronic myeloid (CML)*. http://www.cancer.org/docroot/CRI/CRI_2x.asp?sitearea=&dt=83 (accessed March 4, 2008).

———. 2007b. *Cancer facts and figures*. Atlanta, GA: American Cancer Society.

———. 2007c. *Detailed guide: Kidney cancer—what is kidney cancer (renal cell carcinoma)?* http://www.cancer.org/docroot/CRI/content/CRI_2_4_1X_What_is_kidney_cancer_22.asp?sitearea= (accessed March 4, 2008).

———. 2007d. *Detailed guide: Leukemia—acute lymphocytic (ALL)*. http://www.cancer.org/docroot/CRI/CRI_2_3x.asp?dt=57 (accessed March 3, 2008).

———. 2007e. *Detailed guide: Leukemia—acute myeloid (AML)*. http://www.cancer.org/docroot/CRI/CRI_2_3x.asp?dt=82 (accessed March 3, 2008).

———. 2007f. *Detailed guide: Leukemia—chronic lymphocytic (CLL)*. http://www.cancer.org/docroot/CRI/CRI_2_3x.asp?dt=62 (accessed March 3, 2008).

———. 2007g. *Detailed guide: Lung cancer—small cell. What is small cell lung cancer?* http://www.cancer.org/docroot/CRI/content/CRI_2_4_1X_What_is_small_cell_lung_cancer.asp?sitearea= (accessed February 27, 2008).

———. 2007h. *Detailed guide: Testicular cancer*. http://www.cancer.org/docroot/CRI/CRI_2_3x.asp?dt=41 (accessed March 4, 2008).

———. 2007i. *Hodgkin's disease detailed guide*. http://www.cancer.org/docroot/CRI/content/CRI_2_4_7x_CRC_Hodgkin_Disease_PDF.asp (accessed March 4, 2008).

———. 2007j. *Overview: Kidney cancer—what is kidney cancer?* http://www.cancer.org/docroot/CRI/content/CRI_2_2_1X_What_is_kidney_cancer_22.asp?sitearea= (accessed March 4, 2008).

———. 2007k. *Overview: Leukemia—acute myeloid (AML). The different types of leukemia*. http://www.cancer.org/docroot/CRI/content/CRI_2_2_1x_The_Different_Types_of_Leukemia_AML.asp?sitearea= (accessed March 3, 2008).

———. 2007l. *Overview: Prostate cancer—what is prostate cancer?* http://www.cancer.org/docroot/CRI/content/CRI_2_2_1X_What_is_prostate_cancer_36.asp (accessed March 4, 2008).

———. 2007m. *Overview: Stomach cancer—what causes stomach cancer?* http://www.cancer.org/docroot/CRI/content/CRI_2_2_2X_What_causes_stomach_cancer_40.asp?sitearea= (accessed February 27, 2008).

ALA (American Lung Association). 2007a. *Chronic obstructive pulmonary disease (COPD) fact sheet*. http://www.lungusa.org/site/apps/nl/content3.asp?c=dvLUK9O0E&b=2058829&content_id=EE451F66-996B-4C23-874D-BF66586196FF&notoc=1#5 (accessed February 27, 2008).

———. 2007b. *Diseases A to Z.* http://www.lungusa.org/site/pp.asp?c=dvLUK9O0E&b=33316#p (accessed March 4, 2008).

California Birth Defects Monitoring Program. 2006. *Discoveries and data center.* http://www.cbdmp.org/bd_intro.htm (accessed March 4, 2008).

CDC (Centers for Disease Control and Prevention). 2006. *Summary health statistics for U.S. Adults: National health interview survey, 2005.* Hyattsville, MD: U.S. Department of Health and Human Services.

———. 2007a. *Birth defects: Frequently asked questions (FAQs).* http://www.cdc.gov/ncbddd/bd/faq1.htm (accessed March 4, 2008).

———. 2007b. *National Center for Health Statistics: Asthma.* http://www.cdc.gov/nchs/fastats/asthma.htm (accessed February 27, 2008).

———. 2007c. *National Center for Health Statistics: Pneumonia.* http://www.cdc.gov/nchs/fastats/pneumonia.htm (accessed February 27, 2008).

IOM (Institute of Medicine). 2006a. *Amyotrophic lateral sclerosis in veterans: Review of the scientific literature.* Washington, DC: The National Academies Press.

———. 2006b. *Gulf War and health, volume 4: Health effects of serving in the Gulf War.* Washington, DC: The National Academies Press.

March of Dimes. 2006. *Quick references and fact sheets: Birth defects.* http://www.marchofdimes.com/professionals/14332.asp (accessed March 4, 2008).

NCI (National Cancer Institute). 2002a. *Bone cancer: Questions and answers.* http://www.cancer.gov/cancertopics/factsheet/Sites-Types/bone (accessed March 4, 2008).

———. 2002b. *What you need to know about bladder cancer.* http://www.cancer.gov/cancertopics/wyntk/bladder (accessed March 4, 2008).

———. 2003. *What you need to know about leukemia.* http://www.cancer.gov/cancertopics/wyntk/leukemia/page3 (accessed March 4, 2008).

NHANES (National Health and Nutrition Examination Survey). 2007. Prevalence of chronic kidney disease and associated risk factors—United States, 1999-2004. *Morbidity and Mortality Weekly Report* 56(8):161-165.

NIDDK (National Institute of Diabetes and Digestive and Kidney Diseases). 2008. *Health information for the public.* http://www2.niddk.nih.gov/ (accessed March 4, 2008).

NRC (National Research Council). 2000. *Scientific frontiers in developmental toxicology and risk assessment.* Washington, DC: National Academy Press.

———. 2001. *Evaluating chemical and other agent exposures for reproductive and developmental toxicity.* Washington, DC: National Academy Press.

Ries, L. G., D. Melbert, M. Krapcho, B. A. Miller, E. J. Freuer, L. Clegg, M. J. Horner, N. Howlader, M. P. Eisner, M. Reichman, and B. K. Edwards. 2007. *SEER cancer statistics review, 1975-2004.* http://seer.cancer.gov/csr/1975_2004/ (accessed March 3, 2008).

WHO (World Health Organization). 2003. *World health report 2003—shaping the future.* http://www.who.int/whr/2003/en/ (accessed February 28, 2007).

7

Cohort Descriptions

This report builds on the findings reported in *Gulf War and Health, Volume 1: Depleted Uranium, Pyridostigmine Bromide, Sarin, Vaccines* (IOM, 2000), hereafter referred to as *Volume 1*. The present chapter describes the published scientific literature on potential health effects of uranium in humans. Three major categories were constructed to organize the relevant studies: those of workers occupationally exposed to uranium in uranium-processing plants, those focusing on depleted-uranium exposure of deployed populations (some of which were exposed to depleted uranium through friendly-fire incidents), and those assessing exposure to uranium from environmental sources, including drinking water. The studies on the occupationally exposed workers generally had better study design and methods, especially for assessing exposure to uranium and disease outcomes. Studies of deployed populations, in contrast, had limited or no exposure data other than data on deployment itself and on the possibility of exposure to depleted uranium, but the committee chose to include these studies because of their relevance to the Gulf War and Operation Iraqi Freedom veterans. The chapter also includes studies that assessed health outcomes in people who lived near uranium-processing facilities or had high concentrations of uranium in their drinking water; these residents may have exposure conditions similar with veterans who received level III exposures (see Chapter 5).

The chapter first presents an overview of the cohort studies of processing workers examined in Volume 1 and summaries of derivative studies published after that report. A summary of new cohorts introduced into the literature since 2000 follows. The chapter then describes studies that assessed mortality patterns and health outcomes in deployed service personnel. It ends with studies of environmental exposures. In each instance, the study populations and methods—including study design, measures of exposure, and assessment of outcomes—used by the investigators are described. Tables that summarize the studies are included at the end of the chapter.

The traditional 5% level of statistical significance is used in reporting findings. Results that did not reach the 5% level of statistical significance are described as nonsignificant.

URANIUM-PROCESSING COHORTS

Studies of workers in the uranium-processing and uranium-machining industry are essential for understanding the long-term health effects of uranium exposure. Cohort studies assessing mortality patterns in processing workers have been conducted for some time. The studies of interest include uranium millers and other processors working in plants that process and refine uranium ore into metals for commercial and nuclear use. During refinement and enrichment, workers are exposed to a number of hazardous substances, including chemical toxicants and potential carcinogens. Processing workers are exposed primarily to uranium oxides and derivative uranium compounds produced during the refinement process and other substances that contribute to adverse health outcomes. The studies described present a picture of diverse work histories and varied levels of exposure to enriched uranium, soluble and insoluble uranium compounds, other radioactive elements (such as radium and thorium), and other potentially hazardous industrial chemicals (such as sulfuric acid and fluorocarbons).

In occupational settings, exposure is often prolonged, occurring over a period of several months to years in contrast with the shorter periods of exposure experienced by Gulf War veterans in friendly-fire incidents. Exposure also was greatest in the early years of the procurement and processing initiative in the United States, when safety measures were not as stringent. In occupational studies, exposure is generally assessed through work histories using cumulative measurement of exposure. Inhalation of dust that contains uranium compounds was the primary route of entry of uranium in processing plants, a route analogous to that of many Gulf War veterans exposed to depleted uranium during friendly-fire incidents.

This section first details the cohorts reviewed in *Volume 1*, including updates on the cohorts published after the release of the report in 2000. That information is summarized in Table 7-1. In general, cohorts that did not have updates since that report are not included here, but they are included if there are data on health outcomes in them that were not considered in *Volume 1*. The section then describes new processing cohorts, including studies of uranium processors in the United Kingdom.

Colorado Plateau Uranium-Mill Workers

In assessing long-term health effects of uranium in the processing and machining industry, exposure to uranium and thorium-230 in mill workers is of particular interest (Waxweiler et al., 1983). Before World War II, uranium mining in the Colorado Plateau states was on a relatively small scale; efforts were directed primarily to recovery of vanadium contained in the ore. The establishment of a domestic uranium-procurement program sparked growth and expansion of uranium mining and milling in the United States after the war (Pinkerton et al., 2004). In that effort, uranium mills carried out extraction and purification of uranium ore for commercial use. The enrichment process exposed workers to a number of substances, including dust that contained vanadium, thorium, silica, and radium radionuclides in addition to uranium (IOM, 2000). Concerns about health risks associated with uranium milling were raised as early as 1949, when Colorado health officials submitted a formal request to the US Surgeon General to examine the health of uranium workers (Wagoner et al., 1964). That request, combined with reports from central Europe that documented an increased incidence of pulmonary malignancies in miners and millers, prompted the US Public Health Service to initiate a program to monitor health hazards in the uranium-mining and -milling industries (Wagoner et al., 1964; Archer et al., 1973). As radon exposure in the mines emerged as the primary health issue, the mortality experience of mill workers received little attention. As a result, there was little information on hazards in the uranium-milling industry. The studies described below sought to investigate the potential health effects associated with uranium milling in workers in the Colorado Plateau region.[1]

Wagoner et al., 1964

In the earliest study of the Colorado Plateau mill workers, Wagoner and colleagues assessed mortality in 5,370 white male uranium miners and mill workers. The study population included three subcohorts, one consisting of 611 millers with no reported mining experience. The workers were prospectively identified and had volunteered for at least one physical examination.

Followup of the cohort included triennial physical examinations, an annual uranium-mining industry census, and collection of personal data through mail and other methods. Census takers conducted annual interviews with mine and mill workers, and correspondence was sent to those who could be located. Vital status of 95% of the study group through December 31, 1962, was determined, and mortality was compared with that in the male population of the Colorado Plateau states. Death certificates for 317 workers known to be deceased were obtained

[1]The Colorado Plateau region includes Arizona, Colorado, New Mexico, and Utah.

and classified according to the sixth edition of the *International Classification of Diseases* (ICD-6). The authors examined a number of health outcomes, including cancers of the digestive, respiratory, and lymphatic systems; nonmalignant respiratory diseases; and cardiovascular-renal disease. Workers were categorized by duration and type of employment (milling or mining) through July 1960 to ascertain exposure. They were initially assigned to three cohorts on the basis of type of work experience. Radiation exposure of a small group of miners was computed on the basis of months of underground experience, estimated dose rate, and cumulative dose. Milling experience of workers with mixed industry experience was discounted, and person-years starting with the date that mining experience began were added. The modified life-table technique was used in the analysis of mortality, and age-, race-, and cause-specific mortality was compared with that in the male population of the Colorado Plateau area by using standardized mortality ratios (SMRs).

The authors found no significant differences in deaths between the mill subcohort and the general population (56 observed vs 55.8 expected; SMR, 100). Deaths from all forms of cancer were fewer than expected (6 observed vs 8 expected; SMR, 75). However, there was a slight excess of deaths from cardiovascular-renal disease (28 observed vs 25.3 expected).

The strengths of this study include a well-defined cohort, inclusion of smoking history, and a sound method of followup through annual interviews. In addition, the mortality analysis was conducted by using a local reference population rather than the US population. Despite efforts to measure exposure on an individual level, biologic monitoring was not carried out; exposure was represented only by work classification. Furthermore, exposure assessment was limited by the lack of specificity of exposure because of the possibility of concomitant exposure, and the cohort was relatively small.

Archer et al., 1973

In a second prospective cohort study of Colorado Plateau mill workers, Archer and colleagues extended the followup of mortality in mill workers with a group of 662 white male millers. The men had worked in one of six mills during 1950-1953 and were available for medical examinations (including blood and urine tests, a physical examination, and chest films) in 1950, 1951, and 1953. Occupational and social histories were also documented by the authors. Vital status through December 31, 1967, was ascertained through Social Security Administration (SSA) records and several other sources. Death certificates were collected, and underlying causes of death were coded according to ICD-6. Only 1% of the study population was lost to followup. Person-years by 5-year age group and calendar year were calculated with the modified life-table technique, and cause-specific mortality was compared with that in the white male population in the Colorado Plateau region (Colorado, Utah, New Mexico, and Arizona) by using SMRs.

Overall mortality and mortality from major cardiovascular diseases were consistent with the numbers of deaths expected (104 observed vs 105.11 expected and 52 observed vs 47.72 expected, respectively). Mortality from all other causes was significantly less than expected (12 observed vs 22.42 expected; SMR, 54; $p < 0.05$). However, there was a nonsignificant excess in the number of deaths due to all cancers (20 observed vs 18.11 expected; SMR, 110). A significant excess in deaths from lymphatic and hematopoietic malignancies other than leukemia was also noted (4 observed vs 1.02 expected; SMR, 392; $p < 0.05$). None of the four people who died had worked in or around the furnace area where exposure to uranium and vanadium was greatest.

The study exhibited many of the same strengths as the previous study by Wagoner and colleagues (1964), including the use of a regional comparison group in the mortality analysis and a small sample (662). In contrast with Wagoner et al., Archer and colleagues did not conduct annual followup of the study group.

Waxweiler et al., 1983

In this retrospective cohort study, the authors continued through 1977 followup of the Colorado Plateau mill-worker cohort first studied by Wagoner and colleagues (1964). Microfilmed personnel records of workers in seven uranium mills were used in the selection of 2,002 men who were employed at least 1 day after January 1, 1940; had worked for at least 1 year in uranium mills; and had no work experience in an underground uranium mine. Demographic data and work histories through 1971 were collected and coded for study use.

Researchers found that about half the cohort was employed before 1950 and that only a small number had worked longer than 5 years. Vital status was determined through SSA records and other sources for all but 2% of the cohort. Death certificates were obtained for 515 (97%) of the 533 deceased workers. A total of 43,252 person-years were observed for the study group. Mortality was analyzed with the National Institute for Occupational Safety and Health (NIOSH) modified Life Table Analysis System (LTAS), and cause-, age-, race-, sex-, and calendar-period-specific SMRs were calculated. Mortality was compared with national rates. The authors sought to assess the association between mill employment and lymphatic and pulmonary malignancies, toxic effects in the kidneys due to uranium exposure, and all-cause mortality.

There were statistically significant deficits in total mortality (533 observed vs 605.2 expected; SMR, 88; 95% CI, 81-96) and mortality from all malignant neoplasms (82 observed vs 109.8 expected; SMR, 75; 95%, CI 59-93). There were also fewer deaths due to lung cancer than expected (26 observed vs 31.4 expected; SMR, 83; 95% CI, 54-121), but this difference did not reach the 5% level of statistical significance. The authors reported a statistically nonsignificant excess in mortality from Hodgkin lymphoma (SMR, 231; 95% CI, 48-675). Of the nonmalignant outcomes, there was a significant excess in deaths due to nonmalignant respiratory disease (SMR, 163; 95% CI, 123-212) as a result of increases in the "other nonmalignant respiratory diseases" category. The authors also observed a nonsignificant increase in deaths due to chronic renal disease (SMR, 167; 95% CI, 60-353), but in all cases employment in the mill was rather brief and induction time ranged from 4 to 19 years. Moreover, death certificates in at least half of the six cases also indicated either prostatic obstruction or prostatic cancer.

Pinkerton et al., 2004

Pinkerton and colleagues examined mortality in 1,484 men who worked in one of the seven uranium mills in the Colorado Plateau area. The cohort was drawn from personnel records of uranium-mill workers previously described by Waxweiler and colleagues (1983) that included 2,002 men employed at least 1 day after January 1, 1940, worked for at least 1 year, and had never worked in an underground uranium mine. The authors reviewed records from the Waxweiler et al. study to ensure the inclusion of workers who met the original criteria but were omitted from the study, and they recoded all work histories to remedy any inaccuracies. The resulting subcohort of 1,485 included men who satisfied the original cohort criteria, had never worked in an aboveground or underground mine, and had worked for at least 1 year in the seven uranium mills at the time of uranium or vanadium concentrate recovery. One person was excluded from analysis because his work history was incomplete.

Of the 1,485 men, 97% (1,438) were members of the original Waxweiler et al. cohort. All workers included in the study were followed through the end of 1998. Vital status was determined on the basis of records of SSA, the Internal Revenue Service (IRS), the US Postal Service, and state bureaus of motor vehicles. Death certificates and data from the National Death Index (NDI) were collected to ascertain cause of death of 794 (98%) of the workers known to be deceased and recoded in accord with ICD-9. Less than 2% of the study population had been lost to followup. The End Stage Renal Disease Program Management and Medical Information System (ESRD) was used to assess the risk of death from ESRD and renal disease in the cohort.

The NIOSH modified LTAS was used in the analysis of mortality. The number of person-years at risk was calculated and stratified into 5-year intervals by age and calendar time. SMRs were calculated by using US adjusted mortality and observed deaths. In addition, mortality was stratified by duration of employment, time since first employment (latency), and year of first employment. Mortality was also compared with that in the Colorado Plateau states.

In general, mortality from all causes and mortality from all malignant neoplasms were less than expected in comparison with the US population. Cancer mortality was consistent with findings reported in previous studies of this cohort

(Archer et al., 1973; Waxweiler et al., 1983). Mortality from tracheal, bronchial, and lung cancer for the first time exhibited a statistically nonsignificant increase (78 observed vs 68.93 expected; SMR, 113; 95% CI, 89-141) that was not found in earlier studies of this cohort. Mortality from tracheal, bronchial, and lung cancer was higher in those employed before 1955 (SMR, 134; 95% CI, 102-174) than in those hired in 1955 or later (SMR, 79; 95% CI, 49-121), but a reverse association was observed between tracheal-, bronchial-, and lung-cancer mortality and duration of employment (that is, longer employment was associated with lower mortality). The excess based on regional rates (75 observed vs 49.73 expected; SMR, 151; 95% CI, 119-189) was statistically significant and greater than the excess based on US rates since 1960.

Pinkerton and colleagues also reported nonsignificant increases in deaths from lymphatic and hematopoietic cancers other than leukemia (16 observed vs 11.08 expected; SMR, 144; 95% CI, 83-235) and from chronic renal disease (8 observed vs 5.91 expected; SMR, 135; 95% CI, 58-267). The increase in lymphatic cancers was less than the excess observed by Archer et al. (1973) (SMR, 392; 95% CI, 194-590), which reflected an excess of deaths from Hodgkin lymphoma and lymphosarcoma and reticulosarcoma, but greater than that found by Waxweiler et al. (1983) (SMR, 119; 95% CI, 21-217). In addition, significantly fewer deaths from all digestive cancers than expected were observed (SMR, 62; 95% CI, 43-87).

The authors also observed a significant increase in mortality from nonmalignant respiratory disease (SMR, 143; 95% CI, 116-173) that was due to a significant excess in mortality from emphysema (SMR, 196; 95% CI, 121-299) and pneumoconiosis and other respiratory disease (SMR, 168; 95% CI, 126-221). Mortality from emphysema was higher in workers employed before 1955, when exposure to uranium, silica, and vanadium was thought to be highest (17 observed; SMR, 222; 95% CI, 129-356), than in those employed in 1955 or later (4 observed; SMR, 130; 95% CI, 36-333). However, there were no corresponding differences in mortality from pneumoconiosis and other respiratory diseases. Fewer deaths from nonmalignant digestive disease than expected were observed (SMR, 62; 95% CI, 39-94).

The strengths of this study include a long followup period (vital status through 1998 was determined) and the use of ESRD data that provide useful details on mortality from chronic renal disease associated with uranium milling. Like previous studies of the cohort, this one lacked assessment of individual exposure to uranium and other substances in the milling environment.

Fernald Feed Materials Production Center Workers

From 1951 to 1989, the Fernald Feed Materials Production Center (FFMPC) in Ohio chemically processed uranium-ore concentrate and uranium of low enrichment grade into uranium-metal products. The process involved the use of hydrofluoric acid, ammonia, nitric acid, sulfuric acid, tributyl phosphate, trichloroethylene (TCE), and cutting fluids through dissolution, evaporation, and denitration to produce pure uranium metal. During operation, the facility monitored internal and external radiation exposure.

The Comprehensive Epidemiology Data Resource (CEDR) created by the Department of Energy (DOE) warehoused information regarding workers at multiple nuclear processing facilities, including FFMPC, for 30 years.

Boiano et al., 1989

In 1985, NIOSH investigators conducted a cross-sectional medical assessment of workers at FFMPC. The study followed requests for investigations due to concerns about potential exposure to reported releases of uranium oxides from dust collectors in November and December 1984. About 850 hourly workers were employed at FFMPC at the time of the evaluation. The study focused on evaluating hazards related to lung and renal toxicity after consultation with plant management and workers.

Of the 208 eligible long-term employees (147 hourly and 61 salaried), 146 (70%), identified through employee rosters, participated in the study. The study population consisted of 142 men and four women with mean age and median age of 56 and 58 years, respectively. The employees had worked at the facility for 10-34 years (median, 32 years). Study participants included hourly and salaried employees who had worked at the facility continuously for 10 years, salaried employees who had previously been hourly and who had worked there for at least 10 years, and former hourly employees who had retired within the preceding 2 years after working there continuously for at least 10 years.

A medical and occupational questionnaire was carried out with tests that included blood and urine analysis, chest radiography, and pulmonary-function tests (standard screening spirometry). The questionnaire was self-administered and collected details on an employee's medical history with emphasis on respiratory and renal conditions, occupational and job exposure history, cigarette and alcohol use, and basic demographic information. Questions on respiratory conditions were extracted from the American Thoracic Society questionnaire. On the basis of responses to relevant questionnaire items, the investigators categorized breathlessness in five grade levels; 1-second forced expiratory volume (FEV_1) and forced vital capacity (FVC) served as indicators of pulmonary function. Blood and urine samples were collected for a number of glomerular and tubular biomarkers.[2] Blood and urine tests were used as dichotomous variables in evaluating renal effects. Urinary uranium concentration was also measured.

[2]The markers included beta-2-microglobulin, retinol-binding protein, albumin, total protein, creatinine, *N*-acetyl glucosaminidase, gamma glutamyl transpeptidase, and alanine aminopeptidase.

Personnel records were used to establish work histories, and high-, medium-, low-, and no-exposure groups were created on the basis of job titles. Urinary-uranium data were used to construct exposure histories; the data were used to verify exposure groups and to weight the exposure categories. A cumulative uranium-exposure index was created for each participant by multiplying duration of employment and the potential for uranium exposure in the job held. Radiation measurements (whole-body radiation counts) were used to determine uranium lung burden.

Urinary uranium concentrations varied up to 13 μg/L, and 109 (92%) of the participants had concentrations under the detection limit of 5 μg/L. No associations between glomerular and tubular markers and measures of uranium exposure were observed. The ratio of FEV_1 to FVC was associated with the job-history–derived uranium-exposure index after adjustment for smoking. Shortness of breath was significantly associated with self-reported uranium-exposure incidents.

Ritz, 1999

A cohort of workers employed at FFMPC during its period of operation (1951-1989) was assembled through CEDR (Ritz, 1999). Some 4,014 white male workers were identified; most of them were employed before 1960. The cohort was followed through 1979 with SSA records and from 1979 to 1989 with the NDI to determine vital status; 1,064 had died, and death certificates were obtained.

Internal exposure to uranium was measured through urinalysis and extrapolation from environmental sampling, and external exposure was measured with film badges. Internal exposure, measured as annual lung dose, was mostly exposure to various insoluble forms of uranium, from depleted through enriched in ^{235}U. Because urinalysis reflects inhalation and internal transport of soluble uranium, air sampling was incorporated to provide a rough measure of the risk of inhalation of insoluble uranium as well. Internal exposure from radionuclides was responsible for the bulk of the radiation doses recorded; most monitored workers (68.9%) received cumulative external radiation doses of less than 10 mSv, only 2.6% had doses of over 100 mSv, and none exceeded 300 mSv.

Because CEDR maintained records of exposure to TCE and cutting fluids, these were controlled, as was socioeconomic status (SES). SMRs were calculated by using Monson's life-table analysis. The numbers were too small to conduct separate dose-response analyses for specific cancers, so only the organ systems with the highest likelihood of exposure—respiratory, transport (blood and lymph), excretory, and upper gastrointestinal—were examined individually. Cumulative dose was lagged by 0, 10, and 15 years to allow for cancer latency. The risk-set approach (similar to the nested case-control approach) of Breslow and Day was used for dose-response comparisons; each cancer death was matched to all cohort members who were still alive at the calendar time of the index subject's death (on average 3,300 survivors per death).

All-cause mortality was lower in the workers than in the US white male population (SMR, 84; 95% CI, 79-89). Risk of death from all malignant neoplasms was nonsignificantly higher than in the general population (SMR, 109; 95% CI, 98-122) and lung-cancer mortality was similar (SMR, 101; 95% CI, 83-121). External radiation doses in excess of 100 mSv (which occurred in only 2.6% of the cohort) increased mortality from all cancers, all radiosensitive cancers, and lung cancer, but the numbers were too small for precise estimates (all were nonsignificant). Cumulative external radiation showed a dose-response relationship when lagged by 10 or 15 years and adjusted for chemical exposure and internal dose for all cancers (10-year lag: rate ratio [RR], 1.79; 95% CI, 1.12-2.86; 15-year lag: RR, 1.92; 95% CI, 1.11-3.32), all radiosensitive cancers (10-year lag: RR, 1.88; 95% CI, 1.06-3.32; 15-year lag: RR, 2.0; 95% CI, 1.02-3.94), and lung cancer (10-year lag: RR, 2.13; 95% CI, 1.08-4.18; 15-year lag: RR, 2.77; 95% CI, 1.29-5.95), but not for hematopoietic and lymphopoietic cancers.

Cardiovascular mortality was lower in the workers than in the US white male population (SMR, 78; 95% CI, 71-86), as was emphysema mortality (SMR, 21; 95% CI, 4-60). Differences in mortality from all other causes were nonsignificant. The healthy-worker effect probably accounts for those lower SMRs. The SMR for all-cause mortality remained lower when Fernald workers were compared with NIOSH-Computerized Occupational Referent Population System (CORPS) workers (SMR, 81; 95% CI, 76-86); the comparison should have reduced bias caused by the healthy-worker effect.

The strengths of this study are that it used one of the largest cohorts with monitored external and internal exposures at the individual level, had a long followup period, allowed for a lag period for development of radiation-related solid tumors, and adjusted for other exposures, such as to TCE and cutting fluid, and for socioeconomic status.

Oak Ridge Nuclear Facilities Workers

Oak Ridge, Tennessee, was the home of several nuclear facilities involved in nuclear-weapons production during World War II. Two of the facilities (Y-12 and K-25) were dedicated to uranium enrichment for use in atomic weapons, and a third (X-10, also called Clinton Laboratories) was an experimental laboratory designed to produce plutonium for further research. The Y-12 uranium-processing plant was run by the Tennessee Eastman Corporation TEC) in 1943-1947. With an electromagnetic separation process, uranium was enriched in ^{235}U for use in atomic weapons. In 1947, ownership passed to Union Carbide Corporation, and the plant shifted toward weapons fabrication and research and development for the separation of isotopes. K-25 was a gaseous-diffusion plant for producing enriched uranium.

Polednak and Frome, 1981

In the 4 years that TEC operated the Y-12 plant, about 45,000 people worked there; 38,000 left the plant when Union Carbide assumed management. Demographic and payroll data were submitted to SSA in 1974 for validation of decedents. In 1981, a subset of the 38,000 was examined, and mortality assessed (Polednak and Frome, 1981). The researchers excluded women (47%) from the sample because of incomplete recordkeeping at SSA. An additional 307 temporary workers (those who worked less than 2 days) were excluded, as were 543 because of inaccurate or incomplete data at SSA or the plant. Minority-group members were also excluded (no reason given), leaving a total of 18,869 white men for analysis. Death certificates through 1977 were verified, and all underlying causes of death recorded on the certificates (according to ICD-8 classification) were recorded for analysis.

Electromagnetic separation involved the use of mass spectrograph units in a two-step enrichment process. Both steps exposed workers to uranium dust but little external radiation; the exposure that posed a risk was inhalation of radioactive compounds. Air-sampling records obtained from TEC showed higher than normal readings at concentrations above acceptable limits. In 1945, the first step was eliminated, and this reduced levels of uranium dust, but the enrichment was higher. Company records indicate that particle size varied, some particles being smaller than 1 μm. Film badges were rarely used, because exposure to external radiation was low owing to the nature of the operation; respirators were required but might not have been used with great frequency.

Polednak and Frome examined whether mortality was associated with longer periods of employment and exposure to higher levels of airborne uranium. Workers were categorized by stage of enrichment and other job classifications to determine an approximate level of exposure. SMRs were calculated by using both an internal comparison group (workers in buildings where uranium was not being processed) and external comparison group (the US white male population).

SMRs were below 100 for all-cause mortality and all-cancer mortality; most of the results were nonsignificant. After correction for incomplete ascertainment of cause of death (access to death records was possible for only 95% of deaths), mortality from lung cancer had an SMR of 122 (95% CI, 11-136). Considering type and length of employment resulted in no discernable differences among strata.

Checkoway et al., 1988

Researchers examined 6,781 white men who worked at the Y-12 plant for at least 30 days during the period May 31, 1947-December 31, 1974 (Checkoway et al., 1988). They excluded those who had worked at the plant before May 4, 1947 (that is, before it was turned over to Union Carbide), those who had worked at other facilities, and those with unknown employment dates; they also excluded nonwhite workers and women. The median age of cohort members at hire was 27.6 years, and the median duration of employment was 9.2 years.

The cohort was retrospectively followed through 1979, allowing at least 5 years of followup (median followup time, 20.6 years). SSA and Tennessee Motor Vehicles Department records and state death indexes were used to assess vital status, and death certificates were coded according to ICD-8. Vital status of 96% of the cohort was determined, and person-years of missing cohort members were added up to the date of last contact. A total of 133,535 person-years were observed, and death certificates for 846 (98%) of the 862 deaths were obtained.

Monitoring for external radiation began in 1948 with film badges and switched to thermoluminescent dosimeters in the late 1970s. Whole-body dose equivalents were determined with both methods and summed to provide cumulative exposure values. Researchers assumed that periods in which a worker was not monitored were associated with a low likelihood of exposure and assigned a value of zero to them. The mean cumulative external dose for monitored workers was 0.96 rem over a 31-year period. Internal radiation exposure was determined with urinalysis (which began in the early 1950s) and in vivo counting of deposited uranium (which began in 1961). Lung dose equivalents were determined with metabolic models and accounted for level of uranium enrichment. Internal-exposure monitoring data were available on 3,490 workers. The mean internal dose to monitored workers was 8.21 rem.

All 6,781 cohort members were included in the overall analysis of mortality patterns. The dose-response analyses, however, were restricted to 5,278 workers on whom external-dose data were available and 3,490 (51.5%) workers on whom internal-dose data were available.

Cause-specific mortality was compared to US and Tennessee mortality; SMRs were calculated with a modified life-table analysis with adjustment for age and calendar year in 5-year intervals. Poisson regression analysis was used to determine RRs with adjustment for dose category, age, calendar year, and duration of followup. An internal referent category was used that consisted of the lowest cumulative-dose category for both internal and external exposure. In addition, 10-year latency was assumed.

No significantly increased SMRs were observed for deaths from any cause except lung cancer in comparison with the US population (89 observed; SMR, 136; 95% CI, 109-167). Dose-response trends were detected for lung-cancer mortality with respect to cumulative alpha and gamma radiation. Nonsignificant excesses were observed for cancers of the brain and central nervous system, kidney, and lymphatic system.

Frome et al., 1990

Workers employed at the Oak Ridge facilities during World War II were selected for long-term health-status followup (Frome et al., 1990). The cohort was defined as white men who worked at the facilities for at least 30 days from the initial date of operation through December 31, 1947. Information was found on 28,008 of those who met the criteria.

Monitoring for radiation exposure was infrequent during the period of employment, so exposure was categorized on the basis of likelihood. Job codes and titles with a reasonable likelihood of radiation exposure were categorized as "Y" (yes), and those with no discernible reason for exposure as "N" (no). Workers were also categorized according to the facility they worked at, SES, length of employment, period of followup, and age. Workers contributed person-years from 1950 to 1979 or until loss to followup. Analyses took two approaches: modified SMR (traditional SMR and SMR-trend analyses over 30 years), which allowed the ratio of death rates in the index and reference groups to change over time; and multivariate analyses with Poisson regression modeling to account for the effects of multiple factors (birth year, duration of employment, SES, employment facility, period of followup, and radiation exposure level) on cancer mortality simultaneously.

Unadjusted (crude) analyses showed a statistically significant excess in deaths from all causes (SMR, 111); no healthy-worker effect was observed in this cohort. The SMR for all malignant neoplasms was 105 (nonsignificant). Mortality from respiratory neoplasms was significantly increased (SMR, 125), as was mortality from lung cancer (SMR, 127 based on 850 observed deaths vs 667.99 expected). The increase in lung-cancer mortality was observed after 5 years of followup and increased at 1.44% per year. In the adjusted analysis, the strongest predictor of lung-cancer death was SES, and lung-cancer mortality was not associated with exposure to radiation.

The trend statistics (the rate of change in SMR over a 30-year period) for all causes (0.74), all cancers (1.05) and respiratory cancers (1.36), tuberculosis (3.44), benign neoplasms (6.05), circulatory diseases (1.05), respiratory diseases (1.53), and suicide (2.45) are all significant in an upward direction.

Loomis and Wolf, 1996

The cohort studied by Checkoway et al. (1988) was followed for an additional 10 years until 1990 (Loomis and Wolf, 1996). The study population included workers employed for at least 30 days at Y-12 in the period January 1, 1947-December 31, 1974. Those who had worked at the plant before 1947 were also included in the cohort but analyzed separately. Of those who had worked at the plant in 1947 and later, 8,116 were eligible: 6,591 white men, 922 white women, 449 black men, 149 black women, and five men and women of other racial groups. Of those who had worked at the plant before 1947, 2,841 were eligible: 1,764 white men, 562 white women, 85 black men, 69 black women, and one man of another racial group. The total cohort consisted of 10,597 workers. The researchers noted that the cohort was not identical with that of Checkoway et al. (1988), because of corrections in records.

White men, nonwhite men, and women were analyzed separately, and those who worked at the plant before 1947 were also considered separately. SMRs were calculated and adjusted for age and calendar time. Employment was the only measure of exposure. The mortality experience of this cohort was compared with that of the US population.

Lung-cancer mortality was significantly increased in all workers at the Y-12 plant (SMR, 117; 95% CI, 101-134) and in white men (SMR, 120; 95% CI, 104-138). Lung-cancer SMRs were highest in workers hired before 1955 and those with 5-19 years of employment. No other significant excess or deficit of deaths from any other cancers was observed.

Richardson and Wing, 2006

All Y-12 plant workers who were employed for at least 30 days during the period May 1947-December 1974 were assessed for lung-cancer mortality due to ionizing radiation (Richardson and Wing, 2006). For internal radiation exposure, annual lung-dose estimates were based on results of urinalysis through the 1950s and results of in vivo monitoring that was begun in 1961. The degree of enrichment was included in calculations that involved in vivo monitoring. For years in which monitoring was incomplete, dose was estimated on the basis of exposure potential associated with the department or job title or on the basis of comparison with a monitored equivalent period for the same person. In 1961, a plantwide policy for external monitoring with film badges was implemented; in the late 1970s, thermoluminescent dosimeters began to be used. Exposure in years when there was no monitoring was estimated on the basis of a standard exposure potential associated with the department for the period before 1961 and dose estimates for the period after 1960.

Estimates of lung dose from both internal and external radiation during the period 1947-1985 were calculated for each year of employment for each worker in the cohort, and vital status as of 1990 was obtained from SSA records and the NDI. The study focused on lung cancer as an underlying or contributory cause on the basis of ICDA-8 162 (code 162 of the *International Classification of Diseases Adapted for Use in the United States*). Because smoking history was not available for analysis, two proxies were used: nonlung smoking-related cancers and noncancer smoking-related conditions.

The analysis used a nested case-control method that constructed risk sets by matching noncases to cases on the basis of selection date (that is, a control is chosen at the point when he or she has reached the age of death for the matched case). To ensure comparability, controls were also matched to cases on year of birth, sex, race, SES, length of employment as of the selection date, and employment status on the selection date. Because exposure data were available only through 1985, the analysis assumes a 5-year lag. Conditional logistic regression was used to evaluate the association between lung-cancer mortality and radiation exposure.

RRs for lung-cancer mortality were not significant for either internal or external dose. Joint internal-external exposure calculations showed an increase in mortality as dose increased, but they were not statistically significant. External radiation had a higher correlation with lung-cancer mortality than external and internal radiation combined or internal radiation alone.

This study had several strengths, including exclusion of workers who had never been monitored for internal radiation exposure to reduce bias due to misclassification of exposure and selection bias, direct measurement of both external and internal exposure at the individual level, and sound methodology and analytic approach. Limitations of the study include uncertainty in estimating workers' "missing" external and internal radiation doses and the lack of information on smoking (a strong risk factor for lung-cancer mortality). Inclusion of lung cancer as a contributory cause of mortality created a potential for overestimating the association in that lung cancer is a common site of metastasis from other primary cancers.

Frome et al., 1997

Researchers examined 106,020 workers at all four plants at the Oak Ridge site who were employed for at least 30 days in 1943-1985. Workers were categorized by employment at one of the four facilities or were included in a fifth group if they worked at more than one plant. External radiation exposure was estimated for employees on the basis of limited monitoring; internal exposure was categorized in three groups: "eligible for monitoring and monitored," "eligible for monitoring but not monitored," and "not eligible for monitoring." For dose-response analysis, exposure was lagged. Workers were further stratified by length of employment, SES, birth year, and age. Vital status was determined from SSA files, and participants contributed person-years until death or loss to followup. SMRs were generated on the basis of mortality in the US general population. Groups were also compared for mortality for internal comparisons. Finally, dose-response analysis was included but only for white men.

All-cause mortality and all-cancer mortality in white men were similar to those in the US population (SMR, 100 and 98, respectively). None of the cancers of interest resulted in an SMR that was significantly different from 100. Mortality from most nonmalignant diseases was lower than US population rates: diseases of the blood (SMR, 52), nervous system (SMR, 70), circulatory system (SMR, 95), digestive system (SMR, 80), and genitorurinary system (SMR, 83). The only nonmalignant condition with a reported excess of deaths in white men was respiratory disease (SMR, 112).

Mallinckrodt Chemical Workers

From 1943 to 1966, the Mallinckrodt Chemical Company in St. Louis, Missouri, processed large amounts of uranium ore for production of pure uranium tetrafluoride and uranium metal. During that time, the plant also processed imported ore that contained 70 times the uranium content of North American ore. Processing workers in poorly ventilated areas were potentially exposed to internal and external radiation, in addition to a number of toxicants, including possible carcinogens, silica, and sulfuric acid. Using external radiation dose, Dupree-Ellis and colleagues (2000) sought to assess the link between occupational exposure and mortality patterns in Mallinckrodt processing workers. Lung-cancer deaths in this cohort were included in Dupree et al. (1995).

The study population consisted of 2,514 white male processing-plant workers who were employed in Mallinckrodt during 1942-1966. Workers were selected on the basis of plant records, and vital status was ascertained through the SSA, Pension Benefit Information, and NDI databases. The cohort was retrospectively followed through 1993, with a mean followup time of 34.6 years (median, 36 years). Death certificates for 1,012 of the 1,013 who died during followup were obtained. The study excluded 745 workers whose prior exposure to external radiation was minimal (this group included 556 women [race not specified] and 43 nonwhite men).

Person-years were calculated from 30 days after the date of first hire until the earliest of death, loss to followup, or the end of the study. A total of 87,757 person-years were observed. SMRs were calculated by using underlying cause of death, and the mortality experience of the cohort was compared with that of white men in the United States. Dose-response analyses were stratified on age and calendar period and included nonunderlying cancer causes. An internal comparison group and time-dependent cumulative-dose groups (latency, 10 years for solid tumors and 2 years for leukemia) were used for this purpose.

From the middle of 1945, workers were monitored for external radiation exposure with film badges. Individual annual doses were determined from deep dose-equivalent analysis of film badges, and an algorithm was assigned for years on which data were unavailable. The calculated mean cumulative dose was 47.8 mSv (median, 15.3 mSv). The total population dose was measured at 120,063 mSv.

All-cause mortality was significantly lower than expected (SMR, 90; 95% CI, 85-96), probably because of the healthy-worker effect. There was a nonsignificant increase in all-cancer mortality (SMR, 105; 95% CI, 93-117). Several site-specific SMRs were increased, but none reached statistical significance. The authors reported a nonsignificant increase in excess relative risk of renal-cancer death of 10.5 per sievert (90% CI, 0.6-57.4), and observed:expected ratios of cases of renal cancer by dose were as follows: less than 5 mSv, 3:2.4; 5-9 mSv, 0:0.9; 10-19 mSv, 0:1.3; 20-39 mSv, 2:1.4; 40-79 mSv, 1:1.5; 80-159 mSv, 0:1.3; and 160 mSv and higher, 4:1.2. However, a possible excess risk associated with internal radiation, chemical exposure, or chance could not be ruled out. Chronic nephritis was the only nonmalignant outcome associated with an observed excess in mortality (SMR, 188; 95% CI, 75-381), but this effect was nonsignificant.

Workers at Four Uranium-Processing Operations

In this retrospective case-control study, authors combined data from studies (Polednak and Frome, 1981; Cookfair et al., 1983; Dupree et al., 1987; Checkoway et al., 1988) of workers at four DOE uranium-processing or fabrication operations to explore the relationship between uranium-dust exposure and lung-cancer mortality (Dupree et al., 1995). Study subjects included eligible workers at uranium-processing operations that were managed by TEC from 1943 to the middle of 1947 (TEC operations) and from 1947 onwards (Y-12 operations), workers at the Mallinckrodt Chemical Workers Uranium Division (MCW) at two sites in Missouri from 1942 until operations ceased in 1966, and workers at FFMPC, where production activities ran from 1951 to 1989. The MCW and FFMPC sites also processed uranium-ore concentrate into metal.

Study authors identified 787 lung-cancer cases by mortality followup of the employee cohorts through the end of 1982, which allowed followup of at least 30 years for each cohort. Cases included workers who were employed at any of the facilities for at least 183 days and who died before January 1983 and had lung cancer listed on the death certificate. One control was selected for each case and matched on race, sex, date of birth, and hire date within 3 years. Of the 787 workers, 567 were employed at TEC, 142 at Y-12, 27 at MCW, and 51 at FFMPC. Most of the employees were white men (92%); there were 44 white women, 13 black men, and four black women. Data on complete work history, smoking history (never or ever smoked), and SES (first pay code) were collected from employment and occupational radiation-monitoring records. Smoking data were collected on 48% of the cases and 39% of the controls, with 91% of the cases and 75% of the controls identified as smokers. Health physicists used uranium air-monitoring data and other environmental data to estimate individual annual radiation lung doses from deposited uranium. Cumulative internal and external doses were lagged for 10 years, and smoking status and pay code (monthly or nonmonthly) were accounted for.

Cumulative lung doses ranged from 0 to 137 cGy in cases and from 0 to 80 cGy in controls. In general, there was little evidence of a relationship between internal radiation dose and lung-cancer mortality. The authors reported increased risk in those exposed to 25 cGy or more, with an odds ratio (OR) of 2.0; however, the CI was wide (95% CI, 0.20-20.70) and the result was not statistically significant. Dose-response analyses limited to cases hired at the age of 45 years or more showed higher ORs for exposed workers, but no trend was evident.

Portsmouth Uranium Enrichment Facility Workers

Brown and Bloom (1987) conducted a retrospective cohort study to examine causes of death in 5,773 employees of the Portsmouth Uranium Enrichment facility. The subjects were primarily white men who were employed for at least

1 week during the period September 1954-February 1982. The facility, in Pike County, Ohio, used gaseous diffusion processes to enrich uranium up to 98% ^{235}U. The primary chemical of concern at the plant was uranyl fluoride, which is highly soluble and is a known renal toxicant. The plant had conducted routine (monthly) urine bioassays since 1954 of employees who had the potential to be exposed to toxicants. However, uranium is excreted quickly, and results of the bioassays were not useful in identifying cumulative exposure to the chemical. Instead, the results were used to classify jobs and rank departments on the basis of relative potential for exposure to uranium. The authors identified two cohorts: one with the greatest potential uranium exposure and the other with any potential exposure.

Person-years at risk (PYARs) based on each employee's time at the plant until death were stratified by 5-year calendar periods, age groups, length of employment, and time since first employment. PYARs were multiplied by US white male cause-specific mortality to determine expected numbers of deaths. It was also calculated on the basis of Ohio mortality.

Statistically significant ($p < 0.05$) deficits in mortality from all causes (SMR, 68; 95% CI, 62-75) and from diseases of the respiratory system (SMR, 42; 95% CI, 23-70), the nervous system (SMR, 40; 95% CI, 21-68), the circulatory system (SMR, 72; 95% CI, 62-82), and the digestive system (SMR, 54; 95% CI, 32-86) were identified. The SMR for all malignant neoplasms was 85 (95% CI, 71-102). There were nonsignificant increases in mortality from stomach cancer (SMR, 169; 95% CI, 81-310) and lymphatic and hematopoietic cancers (SMR, 146; 95% CI, 92-218). The authors found a nonsignificant increase in stomach-cancer mortality in those with more than 15 years of employment and 15 years of latency.

The study is limited by its relatively short observation period and its lack of exposure information.

Phosphate-Fertilizer Production Workers

Stayner and colleagues (1985) conducted a retrospective cohort mortality study of 3,199 phosphate-fertilizer production employees in Polk County, Florida. The plant, which produced primarily diammonium and dicalcium phosphates, began operation in 1953. From 1953 to 1958, it also recovered uranium from phosphate ore. NIOSH received reports of a number of lung cancers in nonsmoking workers in 1976 and conducted a survey of the facility to determine exposure to uranium. At that time, all samples from the analysis were below occupational standards. NIOSH also began a study to evaluate mortality in workers at the site. Personnel records were obtained and reviewed; however, in most cases, no detailed job-specific information was available except dates of employment and job titles. Vital status was ascertained by using data from SSA, the Florida Department of Motor Vehicles, and IRS.

PYARs for death were calculated by using the date of hire to December 31, 1997, date of death, or date lost to followup. They were calculated for each race, sex, 5-year age group, and calendar period. Expected deaths were calculated for each 5-year age groups, 5-year calendar period, race, sex, and cause of death. SMRs were calculated for each race and sex.

Cause-specific SMRs for all study subjects and race- and sex-specific groups were not significant at $p < 0.05$. However, the authors note that when lung-cancer deaths were stratified by duration of employment and length of followup, they found an excess in black male workers who had over 10 years of employment and followup (3 observed vs 0.73 expected; SMR, 411; $p < 0.05$).

The study is limited by the lack of job-specific information, information on exposure, and relatively short followup (93% of the cohort had less than 20 years), particularly if one considers that the outcome of interest, lung cancer, has a long latency period.

Nuclear-Fuels Fabrication Workers

Hadjimichael and colleagues (1983) conducted a retrospective cohort study to examine mortality and cancer incidence in 4,106 nuclear-fuels fabrication plant workers in Connecticut. The plant's fuel-fabrication process included receiving enriched uranium, fabricating uranium fuel, encapsulating the fuel in a corrosion-resistant metal covering, and assembling it into larger components for reactors. The subjects had been employed at the plant for at least 6 months during 1956-1978. Personnel records were used to determine job classifications (40 job titles were combined into 16 groups on the basis of similarity of industrial exposure in the manufacturing process). Exposure groups were also developed on the basis of discussions with industrial-hygiene and safety personnel and interviews with supervisors and employees. External-exposure information was obtained from film badges worn by all employees who worked in designated radiation-controlled areas; internal exposure was measured with urine bioassays. SMRs and standardized incidence ratios (SIRs) were calculated for each exposure group, and cause-specific SMRs were calculated for all industrial employees. Connecticut rates were used for indirect standardization. Mortality data were obtained from SSA and from the Connecticut Department of Health Services. Connecticut Tumor Registry data were used to assess the incidence of cancer.

The overall cancer incidence in all male employees (industrial and office employees) was significantly lower than expected (SIR, 0.81; 95% CI, 0.65-0.99). However, brain cancer was marginally significantly higher than expected in industrial male employees (SIR, 2.70; 95% CI, 0.99-5.88). SMRs for male industrial employees were significantly lower than expected for all causes (SMR, 0.83; 95% CI, 0.71-0.97) and lower but nonsignificant for all cancer deaths (SMR, 0.88; 95% CI, 0.62-1.20). Significantly more deaths were observed than expected from chronic obstructive pulmonary disease (SMR, 3.03; 95% CI,

1.11-6.59) and from central and peripheral nervous system diseases (SMR, 3.46; 95% CI, 1.26-7.53).

The study had a number of limitations, including an inability to account for multiple exposures of the study population, the lack of detailed smoking information, and the lack of information on exposures that occurred at previous places of employment. The authors noted that the cause-specific ratios may be underestimated because of the inability to account for 7% of death certificates. Similarly, the SIRs may be underestimated inasmuch as some employees were lost to followup or moved outside the catchment area of Connecticut. Finally, followup was short and may not have accounted adequately for cancers that typically have a long latency.

United Kingdom Processors

Studies of UK processors evaluated cancer incidence and disease-related mortality. Investigating cancer incidence has the advantage of capturing data on people who had malignancies that did not result in death.

McGeoghegan and Binks, 2001

In 1959, the Chapelcross nuclear plant in Scotland, a gas-cooled reactor plant, began operation. In 1980, British Nuclear Fuels PLC (BNFL) began the production of tritium. Researchers conducted a study of the employees of this plant from its inception through 1995 (McGeoghegan and Binks, 2001). The cohort consisted of 2,628 people ever employed at the Chapelcross site before January 1, 1996; there were 63,967 person-years of followup and a mean followup time of 24 years. Subjects were classified as industrial (hourly) or nonindustrial (salaried), and this served as a proxy for SES. Workers were also divided into radiation and nonradiation workers; radiation workers routinely carried film badges. The collective radiation dose received by the radiation workers was 185.1 person-sieverts, and the mean cumulative external dose was 83.6 mSv. The mean annual dose was 8.7 mSv.

Cancer diagnosis (registration) and vital status of each member of the cohort were determined through the National Health Services Central Register (NHSCR), and death certificates were verified; participants contributed person-years until the date of death, the beginning of missing status, or emigration. Age-, sex-, and calendar-year-specific SMRs and SRRs were calculated on the basis of national population statistics for England and Wales and for Scotland obtained from the Office for National Statistics (ONS) and the Information and Statistics Division, respectively. Because Scotland's population is smaller, the Scottish comparisons are not as precise or robust as the English and Welsh comparisons. Adjustments were also made for industrial status (a proxy for SES), worker status, year of joining, length of exposure, length of service, and length of followup. RRs were calculated to determine differences between radiation and nonradiation workers.

Mortality from benign and unspecified neoplasms were significantly higher when compared to English and Scottish populations for radiation workers (SMR, 348) and all workers (SMR, 348), and p values were below 0.05. Endocrine and nutritional diseases, respiratory system diseases, and bronchitis all had higher incidences in radiation workers and all workers than compared to English and Scottish populations, but the SMRs were below 100.

No RRs comparing radiation with nonradiation workers were significantly different from 1.0. For the trend analyses of cumulative external exposure and lag, there was a significant increase in RRs only for bronchitis, with p values below 0.05.

McGeoghegan and Binks, 2000b

The Springfields site at BNFL in Lancashire, UK, was originally a poison-gas factory; in 1948, it was converted to the production of uranium metals. The plant used a chemical separation process to convert yellowcake either to uranium metal or to uranium hexafluoride for further enrichment to uranium oxide fuel.

All 19,589 employees (72% of whom were radiation workers) of the plant were studied; they contributed 479,146 person-years of followup through 1995 (2000b). The vital status of each member of the cohort was determined through the NHSCR, and death certificates were verified.

Age-, sex-, and calendar-year-specific SMRs and SRRs were calculated on the basis of national population statistics and those for the Lancashire area obtained from ONS. Adjustments were made for industrial status (a proxy for SES), worker status, year of joining, length of exposure, length of service, and length of followup. RRs were calculated to determine differences between radiation and nonradiation workers.

External radiation at the site was measured with film badges. The maximum cumulative dose was 769.3 mSv, and the median was 9.3 mSv; 95% of all individual cumulative doses were found to be less than 89.4 mSv.

Deaths from all causes differed significantly between radiation workers and all workers, but the SMRs were below 100. The RR comparing radiation and nonradiation workers was 0.88 ($p < 0.05$). Deaths from all cancers (SMR, 88) and lung cancer (SMR, 86) were significantly decreased and deaths from uterine cancer (SMR, 201) were significantly increased in all workers. Deaths from smoking-related and respiratory system cancers were significantly lower in radiation workers and all workers with SMRs below 100.

Significant deficits in deaths were observed for a number of cancers, including cancers of the stomach (SMR, 71; $p < 0.01$), colon (SMR, 68; $p < 0.001$), liver and gall bladder (SMR, 53; $p < 0.01$), pancreas (SMR, 67; $p < 0.05$), lung (SMR, 72; $p < 0.001$), prostate (SMR, 79; $p < 0.05$), bladder (SMR, 77; $p < 0.05$), and kidney and ureter (SMR, 59; $p < 0.05$). However, there was an increased incidence of uterine cancer in all workers (SMR, 168; $p < 0.05$) and of testicular cancer in nonradiation workers (SMR, 263; $p < 0.05$).

Regarding noncancer mortality, trend analysis with a lag of 15 or 20 years showed significant ($p = 0.037$ and 0.043) positive results for diseases of the nervous and sense organs and with a time lag of 10, 15, or 20 years showed positive results for cerebrovascular disease ($p = 0.023$, 0.011, and 0.037, respectively). The authors also found significant trend results with a lag of 20 years for diseases of the digestive system ($p = 0.034$), prostatic hyperplasia ($p = 0.042$), and violence and accidents ($p = 0.008$).

McGeoghegan and Binks, 2000a

In 1953, the UK Ministry of Supply opened a gaseous-diffusion plant at Capenhurst for the production of enriched uranium; in 1962, the high-enrichment portion of the plant was closed, but nuclear-fuels production continued; in 1973, a gas-centrifuge process was implemented at a second plant; and in 1982, the diffusion plant was officially closed.

Researchers assembled a cohort of all Capenhurst workers employed from the initiation of the plant through 1995 (McGeoghegan and Binks, 2000a). The cohort consisted of 12,543 employees (26% of whom were radiation workers) who contributed a total of 334,473 person-years and had a mean followup of 26.7 years.

The NHSCR was used to determine the vital status of the study participants, and participants contributed person-years until the date of death, missing status, or emigration. Age-, sex-, and calendar-year-specific SMRs and SRRs were calculated on the basis of national population statistics and those for the Capenhurst area, obtained from ONS. Adjustments were made for industrial status (a proxy for SES), worker status, year of joining, length of exposure, length of service, and length of followup. RRs were calculated to determine differences between radiation and nonradiation workers.

External radiation at the site was measured with film badges. The personnel recorded as being at risk for radiation exposure (classified by job status) had a mean cumulative external whole-body dose of 9.85 mSv.

Deaths from all causes in radiation and nonradiation workers showed significant reductions in risk, with SMRs all below 100. The RR between radiation and nonradiation workers was 0.9 ($p < 0.05$). Mortality from all cancers was significantly lower in radiation workers and in all workers than that of the local population of Capenhurst but not lower than the national (England and Wales) population. Lung-cancer mortality in radiation workers and all workers was also significantly lower than that in the Capenhurst population (SMR, 69 and 85, respectively). Mortality from cancer of the pleura was significantly increased in radiation workers (SMR, 496) and in all workers (SMR, 236).

SMRs for nonmalignant diseases of the respiratory and circulatory systems were significantly lower than 100 for radiation workers, nonradiation workers, and all workers.

Rocketdyne/Atomics International Workers

Ritz et al., 2000

Researchers examined 2,297 nuclear-fuel assembly and disassembly workers who were employed at Rocketdyne/Atomics International (RAI), in California, after 1950. The participants were chosen because they had been part of a monitoring program at the site from 1950 to 1993, and extensive internal radiation-exposure data were available. Followup was conducted until death or December 31, 1994. Vital status was ascertained from company records or through SSA, the NDI, or California vital-statistics files, and underlying and contributory causes of death were examined.

Internal radiation was measured with bioassays, in vivo counting, and whole-body counting for the period 1963-1983. Before that, limited internal monitoring was conducted; and after 1983, all radionuclide operations had ceased. External radiation was estimated from RAI records.

Demographic and lifestyle factors for each employee were obtained from personnel records, and workers were categorized. Pay type (hourly, salaried, or managerial) was used as a surrogate for SES, and job location, employment period, or job title was used as a proxy for chemical exposure. Smoking was adjusted for only in a subgroup of the cohort (658 subjects) on whom detailed information was available and was used to assess potential confounding in the larger cohort. The cohort was followed for an average of 25.4 years. Only 0.7% of the workers received an estimated internal radiation dose to the lung greater than 30 mSv, and slightly more than half the workers had recorded doses of 0 mSv.

Two types of analyses were conducted: a comparison with the general population to determine SMRs, which were calculated on the basis of rates in the US white male population; and dose-response analysis of selected combinations of cancer sites (because of low incidence), which used the risk-set approach. Cumulative doses were lagged by 0, 2, and 10 years and adjusted for external radiation exposure.

Mortality for all causes of death was significantly lower than expected (SMR, 72; 95% CI, 66-80). No SMR for cancer achieved significance. The test for trend was significantly positive for lymphopoietic and hematopoietic cancers ($p = 0.0001$) and upper aerodigestive tract cancers ($p = 0.0001$). Only the upper aerodigestive tract cancers achieved significance when lagged at 0 years ($p = 0.01$), 2 years ($p = 0.01$), and 10 years ($p = 0.04$), but no clear dose-response relationship was observed.

The authors observed significantly fewer deaths from diseases of the circulatory system (SMR, 68; 95% CI, 58-78) and digestive system (SMR, 41; 95% CI, 21-72), and from all external causes (SMR, 62; 95% CI, 43-86) in the RAI cohort. SMRs for the remaining nonmalignant causes were not significant.

A later paper (Boice et al., 2006a) re-examined the methodology of Ritz and colleagues and noted several limitations: the population was small, the low occupational doses limited the analysis, the exposure information was restricted to radiation doses received at Rocketdyne, and the cumulative lung dose from inhaled radionuclides was used as a surrogate of dose to other organs. The authors tried to remedy some of those methodologic limitations in their followup study (see the next section).

Boice et al., 2006a

Boice et al. (2006a) conducted a retrospective cohort mortality study of 5,801 radiation workers employed for at least 6 months during 1948-1999 at the facility. The approach to identifying the study cohort and the dose-reconstruction methodology were described in detail in another study by Boice and colleagues (Boice et al., 2006b). The cohort was identified primarily by using records from the Radiation Health and Safety Department. All available records were reviewed to determine whether a worker was monitored for radiation exposure externally or internally. External radiation exposure was determined annually, and bioassay data on radionuclide intake were estimated for 16 organs or tissues by using the International Commission on Radiological Protection models. Those models also accounted for delayed dissolution of inhaled material in the respiratory tract. Annual radiation doses received before and after employment were obtained from a variety of databases, including those of the Nuclear Regulatory Commission and DOE. The data were combined to estimate organ doses for each worker.

After conducting a dose reconstruction to estimate exposure, Boice and colleagues (Boice et al., 2006a) compared the observed and expected numbers of deaths on the basis of mortality in the general population of California. To determine person-years of followup, they used the date that was 6 months after the first date of radiation monitoring or July 1, 1948, depending on which was later, and the date of death, December 31, 1999, the date of reaching the age of 95 years, or the date when lost to followup, depending on which was first. The average observation period was 27.9 years. The mean dose from external radiation was 13.5 mSv, and the mean lung dose from combined external and internal radiation was 19.0 mSv. The authors conducted internal comparisons in an attempt to minimize bias generated by comparing rates with those in the general population. Relative risks were generated by using Cox proportional-hazard models.

The authors found that mortality from all cancers (SMR, 0.93; 95% CI, 0.84-1.02) and from all leukemia, excluding chronic lymphocytic leukemia (SMR, 1.21; 95% CI, 0.69-1.97), was not significantly increased. Nor was mortality from any other cancer or other causes of death. No dose-response trends (as determined by Cox regression analyses) were found for any cancers. RRs were calculated for exposure at 100 mSv. For all cancers except leukemia, the RR was 1.00 (95% CI, 0.81-1.24); for all leukemia, excluding chronic lymphocytic leukemia, the RR was 1.34 (95% CI, 0.73-2.45).

The strengths of the analysis include the extensive approach to the dose-reconstruction methodology, including the ability to estimate specific organ doses; the nearly complete followup of the workers; and the large comparison group of unexposed workers. Limitations include the small sample, the low doses recorded, and the incomplete availability of workers' smoking history.

Savannah River Plant Workers

The Savannah River Plant in Aiken, South Carolina, has been in production since 1952 and is engaged in a variety of processes: uranium processing; nuclear-fuel fabrication; nuclear-reactor operation; nuclear-reactor overhaul, modification, and maintenance; nuclear-reactor refueling; and nuclear-fuel reprocessing. The occupational radiation dose to workers at this site was 85% external and 15% internal.

In 1988, researchers examined records of employees of the Savannah River Plant who had worked there from 1952 to 1975 (Cragle et al., 1988). They identified 17,922 people with complete data records. When the cohort was limited to white men who had worked for more than 90 consecutive days and were either salaried or hourly, a total of 9,860 subjects remained. Person-years were calculated from the date of first hire plus 90 days. Those with unknown status at the time of the study contributed person-years until the date of last contact.

Study participants were categorized in three ways: type of employment, date of first employment, and duration of employment. Type of employment served as a proxy for SES and contained three groups: salaried, hourly, and combined. Date of first employment stratified the cohort into those who were hired before 1955 and those were hired in or after 1955 to identify a group with the longest followup. Duration of employment served as a surrogate for exposure and included those who were employed for less than 5 years, for 5-14 years, and for more than 14 years.

Mortality was compared with that in the US white male population, adjusted for age and calendar year, and stratified by the three categories described above.

No SMRs for combined hourly and salaried workers were significant. However, mortality from all causes and from all cancers was significantly lower than expected ($p < 0.05$) in hourly workers (all-causes SMR, 8; all-cancers SMR, 72) and in salaried workers (all-causes SMR, 64; all-cancers SMR, 68). There were also significant deficits in respiratory cancer deaths for both hourly and salaried workers and brain and central nervous system cancer deaths in hourly workers:

SMRs were all below 100. All deaths from infective and parasitic diseases and diabetes were significantly lower in hourly but not salaried workers. Deaths from respiratory, gastrointestinal, circulatory, and genitourinary diseases were all significantly increased in both hourly and salaried workers, but SMRs were below 100. External causes of death were significant in salaried workers only, but again SMRs were below 100.

Atomic Weapons Establishment Workers

Mortality in 22,552 employees of the Atomic Weapons Establishment in the UK was studied by Beral et al. (1988). All employees who worked at the Aldermaston, Fort Halstead, Orfordness, Foulness, and Woolwich Common facilities at any time from January 1, 1951, to December 31, 1982, were included in the study. The average followup time was 18.6 years.

Exposure of 9,389 workers to uranium and other radiation sources was measured with dosimeters. The average cumulative whole-body exposure to external radiation was 7.8 mSv. Internal radiation dose was not estimated, because the dose probably would have varied from organ to organ and absorption and deposition of radionuclides are "often difficult to assess from external measurements."

PYARs were calculated from the date of first hire or January 1, 1951, if the worker was recruited before then (records were incomplete before 1951). The data were stratified by age, sex, calendar period, and social class.

Overall mortality in the employees was lower than that in the general population. Mortality in the employees with radiation records was similar to that in other employees. However, after a 10-year lag, mortality from prostatic cancer (RR, 2.23; 95% CI, 1.13-4.56) and mortality from cancers of ill-defined and secondary sites (RR, 2.37; 95% CI, 1.23-4.56) were significantly increased. A significant dose-response trend was also noted for prostatic cancer when uranium exposure and cumulative whole-body exposure to external radiation were monitored.

The study used a relatively large cohort but was limited in that fewer than half the workers had individual monitored exposure and smoking information was lacking.

Egyptian Processors

In this study of uranium workers in Egypt, Shawky and colleagues (2002) monitored external radiation exposure at two uranium-processing sites. The study population consisted of 86 processors at milling, monazite-production, and yellow-cake production locations who handled ores and materials that had high concentrations of naturally occurring radioactive materials. Work histories and descriptions were recorded. Dust monitoring and bioassays were conducted to determine radiation exposure of workers. Hematologic and renal-function measures were assessed in a clinical evaluation in which all study subjects underwent a complete blood count and measurement of serum creatinine and urea, and urinary uranium. Thirteen study subjects provided spot urine samples for urinary-uranium analysis because timed and 24-hour urine samples were difficult to ascertain. Uranium concentration, expressed in micrograms per liter, was measured with laser fluorimetry. Air samples were collected to measure air uranium concentration. Linear regression was used in the analysis.

Mean urinary uranium concentration was 17.8 μg/L in the 13 participants who provided spot urine specimens; urinary uranium ranged from 8-29 μg/L. There was a correlation between urinary uranium and serum creatinine in the 13 specimens, and mean uranium excretion was more than 20 times the occupational-exposure decision level of 0.8 μg/L.

DEPLETED-URANIUM STUDIES

This section describes studies that examined the health outcomes related to exposure to depleted uranium as a result of military deployment; the studies are also summarized in Table 7-2. The literature focuses on veterans deployed to conflicts in the Balkans and the Persian Gulf region. This section begins with a case series of US Gulf War veterans involved in friendly-fire incidents who received fragments of depleted-uranium shrapnel. Next, it summarizes the cohort studies that examined the mortality experience and cancer outcome of UK Gulf War veterans, followed by studies that assessed cancer incidence primarily in European service personnel deployed to the Balkans. Finally, it summarizes a study on workers exposed to depleted uranium at the FFMPC in Ohio.

Baltimore Veterans Affairs Medical Center Studies

Since 1993, the Depleted Uranium Follow-up Program at the Baltimore Veterans Affairs Medical Center (Baltimore VAMC) has sought to provide clinical surveillance of Gulf War veterans exposed to depleted uranium through friendly-fire incidents. Depleted uranium was first used by US and other military during the first Gulf War as material for tank armor and in weaponry (McDiarmid et al., 2004). During the course of that conflict, soldiers in or on vehicles and tanks "were mistakenly fired on and struck by munitions containing DU [depleted uranium]" (McDiarmid et al., 2000) and are thought to have inhaled or ingested airborne depleted-uranium particles or to have experienced wound contamination by depleted uranium. In addition, some soldiers had multiple tiny fragments of depleted uranium scattered throughout muscle and soft tissue. As a result, the Department of Veterans Affairs established a medical surveillance system to determine health effects in depleted-uranium–exposed veterans, evaluate techniques to measure uranium, and assess possible surgical management of shrapnel (McDiarmid, 2007). The results of the surveillance program are detailed in a number of studies by researchers at the Baltimore VAMC.

Since the start of the Depleted Uranium Follow-up Program, investigators have prospectively evaluated 74 of the estimated 100 survivors of the friendly-fire incidents during the Gulf War (McDiarmid et al., 2006). The program has started biologic monitoring of soldiers and veterans of Operation Iraqi Freedom (McDiarmid, 2007).

In 1993-1994, the first group of depleted-uranium–exposed veterans was evaluated by the Baltimore VAMC team. Of the 33 examined, nearly half were confirmed through skeletal examination as having uranium fragments embedded in a number of locations throughout the soft tissue. They also had much higher mean urinary uranium concentrations than those without retained fragments (4.47 vs 0.03 μg/g of creatinine), but no other effect was detectable (McDiarmid et al., 2000). Those veterans were examined every 2 years to assess functioning of the major target organ systems likely to be affected by uranium (primarily the kidneys, the central nervous system, and the reproductive system). The surveillance protocol, based on uranium's known and presumed toxic properties, consists of a detailed questionnaire to document medical history, socioeconomic background, and occupational exposure and extensive laboratory testing that includes hematologic and clinical-chemistry measures, urinalysis, seminal and blood uranium, renal markers, seminal analysis, and reproductive endocrine measures; neurocognitive testing; and chromosomal analysis[3] to test for chromosomal aberrations and hypoxanthine-guanine phosphoribosyl transferase (HPRT) (McDiarmid et al., 2001). A urinary uranium concentration of 0.10 μg/g of creatinine was used as a cutpoint to compare mean values between "high" and "low" uranium-exposure groups.

The studies discussed below examine the health effects of depleted uranium in a group of Gulf War veterans examined at the Baltimore VAMC in 1997, 1999, 2001, 2003, and 2005. The studies concerned primarily people who had retained fragments of depleted-uranium shrapnel and those who suffered inhalation exposure. Results of the 1997 evaluation were discussed in *Volume 1*; a brief summary of that study is provided.

McDiarmid et al., 2000

Of the 33 Gulf War veterans with retained fragments of depleted-uranium shrapnel examined in 1993-1994, 29 were re-examined in 1997 for clinical health effects associated with friendly-fire exposure, and results were compared with those of examinations of 38 veterans not exposed to depleted uranium. Age and military rank were used to recruit unexposed veterans. The authors used several sources, including Army and National Guard units, advertisements, and a Department of Defense hospitalization database. Exposure status was determined from

[3]Human genotoxic outcomes have been explored in greater detail in Chapter 4; therefore, little attention is given here to measures used in the Baltimore VAMC surveillance.

medical records, telephone screening, and a series of questions about military experience.

Clinical evaluation included a complete history and physical examination and a series of laboratory tests to assess hematologic and renal-function measures. Urinary and seminal uranium concentrations and whole-body radiation counting were used to determine exposure. For total-uranium analysis, 24-hour and spot urine samples were collected. The resulting values were expressed in micrograms per gram of creatinine. Kinetic phosphorescence analysis was used to measure seminal uranium concentrations. The authors measured a number of clinical elements in relation to urinary and seminal uranium concentrations. Traditional (paper and pencil) and automated neurocognitive testing batteries[4] were used, and two impairment indexes (one based on traditional measures and one on automated measures) were created for analysis. The ratio of the total number of below-expectation scores to the scores obtained for each battery was used to determine impairment indexes. When t scores were not available, decision cutpoints were used. A number of reproductive-health measures were analyzed by using concentrations of follicle-stimulating hormone (FSH), luteinizing hormone (LH), thyroid-stimulating hormone (TSH), free thyroxine, prolactin, and testosterone. In addition, semen characteristics (volume, sperm concentration and total count, and functional measures of motility) were evaluated by using World Health Organization (WHO) criteria for semen normality. Peripheral blood lymphocytes were cultured to examine frequencies of chromosomal aberrations and sister-chromatid exchange.

On the basis of 24-hour and spot urinary-uranium values, the participants were divided into high- and low-exposure groups. The high-exposure group consisted of 14 veterans with urinary uranium greater than 0.10 μg/g of creatinine. The low-exposure group consisted of all subjects with spot urinary uranium of less than 0.10 μg/g of creatinine; 15 depleted-uranium–exposed and 38 unexposed veterans were in this category. Researchers used correlation and regression analyses to evaluate exposure measures, using 24-hour urinary uranium as the primary measure of exposure. Results were stratified at the median to create low- and high-result groups and compared with the results in the low- and high-exposure groups to determine an association between higher and lower median tendencies. In a separate analysis, neurocognitive indexes were modeled as a function of urinary uranium concentration with adjustment for intelligence (the Wide Range Achievement Test 3 Reading, WRAT-3 Reading) and psychiatric status (the Beck Depression Inventory, BDI).

[4]Traditional neurocognitive measures included the Wide Range Achievement Test 3 Reading, the National Adults Reading Test, the California Verbal Learning Test, the Trail Making Test Parts A and B, the Shipley Institute of Living Scale, and the Digit Span, Arithmetic, and Digit Symbol subsets of the Wechsler Adult Intelligence Test-Revised. Automated measures included Automated Neuropsychological Assessment Metrics, the Nonverbal Selective Reminding Test, and the Kay Continuous Performance Test.

Seven years after first exposure, veterans who had retained depleted-uranium shrapnel fragments had higher urinary uranium concentrations than those who did not have retained shrapnel. Urinary uranium ranged from 0.01 to 30.74 μg/g of creatinine in veterans with retained fragments vs 0.01 to 0.05 μg/g of creatinine in unexposed veterans. Renal-function measures (serum creatinine, beta-microglobulin, retinol-binding protein, serum uric acid, urinary creatinine, and urinary protein) were quite similar between exposure groups. No significant differences were observed between high- and low-exposure groups in hematologic, reproductive, and genotoxicity measures. In general, outcome measures in the depleted-uranium–exposed were within normal clinical limits. Results suggested a statistically significant relationship between increased urinary uranium and poor performance on automated neuropsychologic tests regardless of the regression model used (24-hour urinary uranium for depleted-uranium–exposed veterans, $p = 0.01$; and spot urinary uranium for all veterans, $p = 0.01$).

McDiarmid et al., 2001

In March-July 1999, researchers at the Baltimore VAMC continued their assessment of clinical health effects in Gulf War veterans exposed to depleted uranium by friendly fire. The study population consisted of 50 male depleted-uranium–exposed Gulf War veterans who had retained fragments and were excreting uranium at increased rates 8 years after first exposure. Of the 50, 21 had participated in previous Baltimore VAMC studies, and 29 were newly identified by the study team. Using published estimates of mean urinary uranium concentrations in unexposed groups (11-22 ng/L) and upper dietary limits as a result of naturally occurring uranium in groundwater (up to 0.35 μg/L), the authors established low– and high–urinary-uranium groups. Of the 50 veterans, 13 veterans had urinary uranium greater than 0.10 μg/g of creatinine and were therefore in the high-exposure group; the remaining 37 were below the cutpoint of 0.10 μg/g of creatinine. Three of the 29 new participants had urinary uranium greater than 0.10 μg/g of creatinine.

The 1999 clinical assessment replicated the protocol from the previous study and consisted of a laboratory examination to evaluate hematologic and renal-function measures, reproductive function (semen quality and neuroendocrine function), and genotoxic measures and a detailed questionnaire history and complete physical examination. Test batteries used for neurocognitive-performance measures were similar to those used in the 1997 surveillance (McDiarmid et al., 2000). The traditional neuropsychologic-test measures were used to create an index score. Three impairment-index scores were obtained from automated measures based on response accuracy, median response time for correct response, and number of correct responses per minute. Reproductive endocrinologic values included measurements of TSH, free thyroxine, and the hormones previously assessed (prolactin, FSH, LH, and testosterone). Only 44 of the 50 samples were considered in the semen analysis because six veterans were azoospermic. Enzyme treatment was used for 17 samples (12 with low and 5 with high urinary uranium). In the assessment of genotoxicity, cultured peripheral blood lymphocytes were tested for chromosomal abnormalities and sister-chromatid exchange for baseline measurements. The cells were subjected to two concentrations of bleomycin (4 and 8 μg/mL). A number of potential confounders were adjusted for in the regression analysis, including current smoking status and use of psychotropic and antidepressant drugs. Robust regression analysis was used to account for highly influential observations of neurocognitive function in adjusting for intelligence (WRAT-3) and depression (BDI).

Urinary uranium ranged from 0.018 to 39.1 μg/g of creatinine in the depleted-uranium–exposed veterans with retained fragments and from 0.002 to 0.231 μg/g of creatinine in depleted-uranium–exposed veterans without fragments. In general, clinical tests revealed hematologic, renal, and neuroendocrine measures that were within normal limits with slight differences between high– and low–urinary-uranium groups. When veterans were assessed for active medical problems, those in the high-uranium group were found to suffer a higher proportion of injuries than those in the low-uranium group (76.9% vs 45.9%; $p = 0.05$). Hematologic measures had statistically significant differences between exposure groups that were not observed in the previous surveillance. The high–urinary-uranium group had a lower mean lymphocyte count (32% vs 37%; $p = 0.04$), a higher mean neutrophil percentage (55% vs 49%; $p = 0.03$), and a lower mean monocyte percentage (7.6% vs 9.1%; $p = 0.01$). The authors did not detect any clinically important changes in renal function due to depleted-uranium exposure. Urinary creatinine concentration was slightly lower in the high–urinary-uranium group, but the difference only marginally significant.

Results of neurocognitive tests were not consistent with those in past evaluations. The relationship between urinary uranium and performance on automated measures observed in the 1994 and 1997 surveillance appeared to weaken and was only marginally significant when WRAT-3 and the BDI were adjusted for high and low urinary uranium.

There were no statistically significant differences in concentrations of FSH, LH, prolactin, testosterone, TSH, or thyroid hormones between low and high groups. Of the 44 sperm samples included in the analysis, three were designated subnormal—that is, as having values of at least three of the five clinical measures below normal, as defined by the WHO standards. The high–urinary-uranium groups showed increases in mean total sperm count (583.5 ± 106.1 vs 286.6 ± 44.8), total progressive sperm (220.9 ± 44.0 vs 108.2 ± 19.2), and total rapid progressive sperm (155.5 ± 31.1 vs 81.3 ± 15.4), and the differences were significant ($p < 0.02$, 0.03, and 0.04, respectively).

McDiarmid et al., 2002

In this study that revisited their previous results, McDiarmid and colleagues identified 30 new members of the originally exposed group. Urinary uranium concentrations were measured, and correlation analyses were conducted to determine the relationship between excretion measures in the 1994, 1997, and 1999 surveillance groups.

An increase in urinary uranium (24-hour urinary uranium concentrations higher than 0.05 μg/g of creatinine) was observed in four of the 30 newly identified veterans. Urinary uranium showed a high correlation (R-squared [rsq] = 0.8623) between the 1994 assessment and the 1997 assessment. A strong correlation (rsq = 0.8764) was also observed between the 1994 and the 1999 assessments.

McDiarmid et al., 2004

In the third surveillance, 39 Gulf War veterans were examined at the Baltimore VAMC during April-July 2001. Of the 39, eight were identified as new participants, and the remaining 31 had participated in the followup program at least once before. As in earlier studies, the authors investigated a number of clinical outcomes as related to urinary uranium concentrations 10 years after the initial exposure to uranium. In addition to the clinical measures assessed previously (McDiarmid et al., 2000, 2001), the evaluation considered immunologic measures and mutagenic effects related to depleted-uranium exposure by assessing HPRT mutation frequency. The 29 participants with no history in the Depleted Uranium Follow-up Program were also evaluated for posttraumatic stress disorder and substance abuse. The exposure groups were defined on the basis of participants' 2001 urinary-uranium results. Thirteen participants were in the high-exposure category (with concentrations greater than 0.10 μg/g of creatinine), and 26 in the low-exposure category.

Neurocognitive test batteries were similar to those used previously (McDiarmid et al., 2001). Automated measures were selected from the Automated Neuropsychological Assessment Metrics Test Library. The authors constructed four indexes of impairment based on the traditional tests and the automated measures (response accuracy, median response time for correct response, and number of correct responses per minute or throughout). The indexes represent the proportion of scores that fell 1 standard deviation below the mean. The BDI was used to evaluate emotional status. Genotoxic tests were adjusted for potential confounders (age, smoking, exposure to genetic toxicants, and cloning efficiency).

Urinary uranium ranged from 0.001 to 78.125 μg/g of creatinine. The presence of retained depleted-uranium shrapnel appeared to be associated with higher urinary uranium concentration. In addition, most urinary-uranium results were fairly consistent throughout a given person's history in the group.

The percentage of veterans who reported suffering injuries during friendly-fire incidents showed some significant differences between high- and low-exposure groups. Mean values of all hematologic and renal-function markers were within normal clinical limits with few statistically significant differences between high– and low–urinary-uranium groups. Differences in hematocrit (42.59% in the high group and 44.60% in the low group) and hemoglobin (14.79 vs 15.40 g/dL) levels were not observed in the 1997 and 1999 surveillance groups. Renal-function measures showed movement toward decreased protein reabsorption and increased glomerular filtration of protein: serum creatinine concentrations were higher in the low-uranium group (0.85 vs 0.95 mg/dL; p = 0.03); values for urinary retinol-binding protein and total urinary protein concentrations were higher in the high-uranium group (retinol-binding protein, 65.68 vs 46.13 µg/g of creatinine; p = 0.06; and total protein, 78.69 vs 54.63 mg/g of creatinine; p = 0.01, respectively). As in past years, neurocognitive measures did not differ between exposure groups. Overall neuroendocrine function was normal, but mean free thyroxine was higher in the low-uranium group (1.66 vs 1.08 ng/dL)—a result not observed in the 1997 and 1999 surveillances. Semen measures were higher in the high-uranium group, but the differences did not achieve statistical significance. Immunologic measures revealed a higher proportion of CD4+ T cells in the high-uranium group (65.98% vs 60.83%) and a lower proportion of CD8+ T cells (26.55% vs 31.28%). The authors reported a statistically significant difference between groups with respect to changes in mean baseline measurements of chromosomal aberrations (0.01 ± 0.004 in the high-uranium group and 0.003 ± 0.001 in the low-uranium group; p = 0.027). There were no statistically significant differences in HPRT mutation frequencies between exposure groups.

McDiarmid et al., 2006

This evaluation of the friendly-fire group took place 12 years after first exposure to depleted uranium. The authors reported on the same health outcomes examined in the 1999 and 2001 evaluations. They examined 32 Gulf War veterans in April-July 2003 for hematologic and blood-chemistry measures, renal function, neurocognitive function, genotoxic measures, and reproductive neuroendocrine function and semen characteristics. Urinary-uranium results obtained in the 2003 examination were used for group composition, and a concentration of 0.10 µg/g of creatinine served as the cutpoint for high- and low-exposure groups. The high-exposure group consisted of 13 with urinary uranium of at least 0.10 µg/g of creatinine, and the low-exposure group 19 with less than 0.10 µg/g of creatinine. Robust regression and polynomial transformation analysis were applied to account for possible outliers in the linear-regression model for the neurocognitive evaluation.

Results for all health measures were within normal clinical limits. The difference in serum phosphate concentration was the only measurable difference (p = 0.03) between the high-exposure group (3.75 mg/dL) and the low-exposure group (4.11 mg/dL). Higher mean values of semen characteristics were observed in the high–urinary-uranium group. Despite the persistently increased urinary uranium concentrations, no clinical abnormality or dysfunction was observed.

McDiarmid et al., 2007

In the 2005 surveillance, 34 members of the depleted-uranium–exposed Gulf War veteran group were examined 14 years after first exposure. The authors used clinical assessments that had been used in previous evaluations to determine urinary uranium concentrations, renal function, hematologic and blood-chemistry characteristics, neuroendocrine measures, semen quality, genotoxicity, and neurologic function. Fluorescent in situ hybridization assay analysis was carried out to detect low-level chromosomal abnormalities. As in past evaluations, data on urinary uranium were divided into low and high groups on the basis of a cutpoint of 0.10 µg/g of creatinine. In addition, investigators measured cumulative uranium exposure for each participant to account for the duration and intensity of exposure, bearing in mind urinary uranium concentrations for each surveillance visit and the time between measurements. The latter metric had a cutpoint of 10 µg/g of creatinine·years (based on the distribution of data) and resulted in a group composition consistent with the current uranium cutpoint of 0.10 µg/g of creatinine. As in previous surveillance years, robust regression was used to account for outliers. Age and cloning efficiency were adjusted for in the analysis of mean HPRT mutation frequencies.

Urinary uranium concentrations ranged from 0.002 to 44.1 µg/g of creatinine for total 24-hour urinary uranium concentration in participants known to have embedded depleted-uranium shrapnel fragments and specific indicators of depleted uranium at or above 0.10 µg/g of creatinine. The results showed a high correlation between current and cumulative uranium exposure, with an R2 value of 0.827.

Results regarding health outcomes were fairly consistent with past evaluations. There were no statistically significant differences between high and low urinary uranium in hematologic, blood-chemistry, and neuroendocrine measures, and they were generally within normal clinical limits. Mean serum uric acid reached significance ($p = 0.03$) when high and low groups were compared for cumulative uranium exposure. Despite that finding, the values were within the normal clinical range, and the difference was rather small: the high group registered 5.22 mg/dL and the low 6.19 mg/dL. Other renal measures revealed no significant differences regardless of the exposure metric used. Results of neurocognitive testing were similar to those in past years. The authors found no statistically significant differences between exposure groups in all neurocognitive indexes when either exposure metric was used. Mean values of semen characteristics also showed no significant difference; however, values of percent progressive sperm and percent rapid progressive sperm were lower in the high-uranium group ($p = 0.15$ and 0.12, respectively) when the current exposure metric was used.

Unlike many studies, the Baltimore VAMC study examined targeted health outcomes in the study group that might have resulted from continuous exposure from embedded depleted-uranium shrapnel. The study had a long-term and consistent followup, particularly in relation to uranium excretion, which allowed one to see the chronic health effects in this population. In the most recent study (McDiarmid et al., 2007), the use of measures of cumulative exposure and depleted uranium enhanced the specificity of exposure. However, the small and select sample and the absence of a control group limited its ability to detect effects in the various outcomes. In addition, the selection of the urinary-uranium cutpoint of 0.10 μg/g of creatinine was not based on a generally accepted standard for urinary uranium.

UK Gulf War Studies

All UK military personnel who were deployed to the gulf region from September 1990 to June 1991 were evaluated in a retrospective cohort study (Macfarlane et al., 2000, 2003). The cohort consisted of 53,462 service members and an age-, sex-, service-, and rank-matched comparison group of 53,462 service members who were not deployed to the gulf region in the same period. Personnel were linked to the NHSCR to determine cancer diagnosis and vital status, and deaths were coded according to ICD-9. After exclusion of those who had died before the end of the Gulf War, those who had emigrated from the UK during the study period, and those whose vital status could not be determined, the Gulf War–deployed group consisted of 51,721 participants, and the nondeployed group consisted of 50,755 participants.

Initial cancer diagnoses in the registry through July 2002 were included in the analysis, and person-years at risk were calculated from April 1, 1991, to the earliest of either the date of emigration from the UK, the date of death, the date of first diagnosis of cancer, or July 31, 2002. Of the 51,721 deployed to the gulf, 2,092 reported an exposure to depleted uranium.

No excess risk of cancer overall was observed in the Gulf War veterans: there were 270 incident cancers in the Gulf War–deployed and 269 in the nondeployed (incidence rate ratio, 0.99; 95% CI, 0.83-1.17). No excess risk of any site-specific cancers was found, and adjustment for lifestyle factors and other potential confounders did not change the results.

Balkans Studies

Gustavsson et al., 2004

Swedish military and civilian rescue personnel deployed to the Balkans in 1989-1999 were studied to determine whether they had a higher incidence of cancer (Gustavsson et al., 2004). Swedish Armed Forces and Swedish Rescue

Services Agency registries were used to assemble the cohort; most subjects served 6-month missions. Each person was matched to the Swedish Cancer Registry, and 99.9% of the subjects could be followed up; this resulted in a cohort of 8,347 military men, 433 military women, 403 civilian men, and 5 civilian women. Person-time was calculated through the end of followup (1999) or until death, emigration, or cancer diagnosis. SIRs were calculated on the basis of cancer incidence in the general population, and adjustments were made for sex, age (5-year age groups), and period. No measurement or modeling for depleted-uranium exposure was included.

There were 34 incident cases of cancer diagnosed during the followup period compared with 28.1 expected in the cohort (SIR, 120; 95% CI, 90-170). Eight cases of testicular cancer were identified in military men compared with 4.3 expected. The authors reported no statistically significantly increased incidence of cancer but recognized that the followup period was too short to assess the long-term risk of cancer.

Nuccetelli et al., 2005

On the basis of reports of possible depleted-uranium–related cancer risk, the Italian Ministry of Defense examined a large portion of the Italian military deployed to the Balkans during December 1995-January 2001. The cohort consisted of about 40,000 soldiers 20-59 years old who had been deployed at least once in that period and contributed about 80,000 person-years. Cancer incidence was calculated for 5-year age groups for all cancers and specific cancers: Hodgkin lymphoma, non-Hodgkin lymphoma, acute lymphocytic leukemia, and solid tumors. SIRs were calculated on the basis of cancer registries for the general Italian male population. No measurement or modeling for depleted-uranium exposure was included.

The overall incidence of cancer was significantly lower than expected. The only cancer that was significantly increased was Hodgkin lymphoma (SIR, 236; 95% CI, 122-436). A limitation of the study is that the followup period was too short to assess cancer outcomes.

Storm et al., 2006

After reports of increased cancer incidence in military personnel deployed to the Balkans, Danish Defence Health Services and the Danish Cancer Society undertook a study of Danish military (Storm et al., 2006). From January 1992 to December 2001, 15,091 persons were deployed to the Balkans. After exclusion of those deployed to other conflicts, those with errors in their files, and those with a previous diagnosis of a cancer, the cohort contained 14,012 people. The entire cohort was followed through December 2002 or until death, emigration, or loss to followup. SIRs were calculated for the personnel by using corresponding incidences in the Danish population.

No significantly increased SIRs were observed for all cancers or site-specific cancers except bone cancer; of these, there were four cases, three of which occurred in the first year. The SIR for all bone cancers was 600 (95% CI, 160-1,530); if the first year was excluded, it was 170 (95% CI, 0-1,010).

Sumanovic-Glamuzina et al., 2003

During the Bosnian War, civilians were potentially exposed to environmental contaminants that might have resulted in increases in malignant diseases and other adverse health effects. However, the magnitude of exposure and the resulting health outcomes remained unclear. In 2000, researchers sought to assess the prevalence of major congenital malformations in two 1-year cohorts of neonates born immediately after the war (in 1995) and in 2000, 5 years after military activities. The study population included all live-born neonates and stillborn fetuses in the maternity ward of the Mostar University Hospital in western Herzegovina. Malformations were documented by using the EUROCAT Protocol during physical examination of live-born and stillborn neonates. Autopsies were not performed on the stillborn. For the 1995 cohort, data on prenatal and perinatal complications were collected from medical records and interviews with the mothers. Interviews were not conducted for the 2000 cohort. Prevalence was analyzed with consideration of the relevant organ systems, sex, and gestational age, and chi-square tests were conducted. Aborted fetuses were not included in the analysis.

In 1995, 40 of 1,853 neonates had major malformation, a prevalence of 2.16% (95% CI, 1.49-2.82%). In 2000, 33 of 1,463 (2.26%) had major malformations (95% CI, 1.50-3.01%). Anomalies of the cardiovascular and central nervous systems were significantly higher in the 2000 cohort than in the 1995 cohort.

ENVIRONMENTAL-EXPOSURE STUDIES

The studies reviewed below examine health outcomes in persons who lived near uranium-processing facilities or in households in Finland where well water with high uranium content was the primary source of drinking water. The studies are summarized in Table 7-3.

Residential Studies

Bithell and Draper, 1999

Greenham Common US Air Force base in Berkshire in the UK was the site of a B-47 jet fire in 1958. Residents living around the base expressed concern that depleted-uranium contamination from the fire might have resulted in an increased incidence of cancer, particularly leukemia, in the area. In a 1961 declassified document, excess concentrations of plutonium and uranium were modeled on a contour map. The concentrations peaked in a line along the runway and at two points about 2 km from each end of the runway, resulting in a dumbbell shape. The concentrations were determined on the basis of the ratio of ^{235}U to ^{238}U in 26 evergreen leaf samples taken from the area. Bithell and Draper (1999) analyzed the information and determined that some errors had occurred in the environmental modeling. They estimated that there was only a 1% increase in uranium in the area and a 0.1% increase in uranium activity on the basis of estimated level of enrichment.

The researchers then used an existing set of data on the incidence of childhood leukemia and non-Hodgkin lymphoma that was constructed specifically to show the spatial distribution of cancers around nuclear facilities. In 1966-1987 throughout Britain, 11,283 cases were diagnosed; expected numbers of cases were calculated, allowing for various socioeconomic and demographic factors. The researchers looked at cancer incidence within 6 km of the runway, counted the children who resided in that area, and compared the cases with estimated values.

Within 6 km of the study area, 15 cases of leukemia were observed compared with 13.4 cases expected. There was no evidence that the observed cases were closer to the airfield than expected; that is, there was no clustering of cancers around the airfield.

Boice et al., 2003b

Reports of cancer clusters around nuclear facilities have prompted a number of population-based descriptive studies to determine whether cancer mortality has increased. In 2003, researchers examined cancer rates in people who lived around two nuclear-materials processing facilities in Pennsylvania. The two facilities were 3 miles apart on the Kiskiminetas River, which forms the boundary between Armstrong County and Westmoreland County. The Apollo facility, which processed uranium fuel, opened in 1957; the Parks facility, which processed uranium and plutonium, opened in 1960. Uranium processing ceased in 1983, plutonium processing ceased in 1980, and both sites have been decommissioned and decontaminated. Both facilities were in Armstrong County, and Westmoreland County is just across the river; researchers examined cancer-mortality data on the two counties. Most of the population lived within 20 km of both sites, although wind modeling suggested that Armstrong County had a greater risk of exposure.

Six counties were chosen as controls on the basis of similarities in race, geographic density, employment status, poverty level, age, education, mean family income, and population size. Other cancer risk factors, such as diet and smoking, could not be controlled for with this method.

Death rates in 1950-1995 based on National Center for Health Statistics data were collected for each county and categorized by cause, sex, race, and 5-year age group. SMRs were calculated for each county on the basis of national statistics. Rates were also divided into three periods to capture deaths before, during, and after the operation of the plants. RRs were calculated from the SMRs between the study counties and the control counties.

No significant excess in mortality was found in the study counties in comparison with the control counties or the US population for all cancers and cancers of a priori interest (lung, bone, hepatic, and renal cancers). Mortality rate ratios for all malignancies were similar in the three periods—before, during, and after plant operation (RR, 0.96, 0.95, and 0.98, respectively).

Boice et al., 2003a

The researchers who studied the Apollo and Parks sites also surveyed cancer incidence in a smaller region around the facilities. They looked at eight boroughs and municipalities, identifying 16,772 people who lived in the area in 1990 in whom cancer incidence could be determined. They then examined cancer incidence in 1993-1997. The population was categorized by age, sex, and race, and person-years in each category were calculated. Expected numbers of incident cancer cases were determined by multiplying person-years by incidence of specific cancers in Pennsylvania. The researchers then determined cancer incidence in each municipality and constructed SIRs.

No significant SIR was found for all cancers or a number of cancer sites, such as lung (SIR, 88), kidney (SIR, 105), non-Hodgkin lymphoma (SIR, 110), liver (3 observed vs 4.91 expected), and bone (2 observed vs 1.19 expected). However, a statistically significant excess of cervical cancer was observed (SIR, 235; 95% CI, 113-433).

Boice et al., 2003c

A similar study was conducted in Karnes County, Texas, where uranium milling and mining began in 1954 (Boice et al., 2003c). The three mills and 40 in situ mines (also known as solution mining) and surface mines were in operation until the early 1990s and created concern about environmental exposure in the resident population. In 1961, the Texas Department of Health began environmental monitoring; in 1988, it began sampling residential areas (including water supplies, homes, and food items).

Researchers examined cancer mortality in the general population of Karnes County. They chose four control counties in the same region in Texas as comparison areas on the basis of similarities in race, geographic density, percentage employed in manufacturing, poverty level, age, education, mean family income, and population size. Numbers of deaths in each county in 1950-2001 were obtained and categorized by cause, sex, race, and 5-year age group. Expected numbers were calculated for each county and compared with the observed values to determine SMRs. In addition, numbers of deaths were combined in the periods

1965-1979, 1980-1989, and 1990-2001 to allow comparisons before, during, and after milling and mining operation. RRs for cancer mortality were computed between Karnes County and the control counties and for three periods (1965-1979, 1980-1989, and 1990-2001).

Overall, 1,223 cancer deaths occurred compared with 1,392 expected (SMR, 88). There was no statistically significant increase in SMRs for cancers of a priori interest (lung, bone, renal, and hepatic cancer). The only exception was colon and rectal cancer in the early period—before and in the early years of operation.

Pinney et al., 2003

During 1951-1988, the FFMPC, in Fernald, Ohio, processed uranium ore and other uranium feed materials for nuclear-weapons production. Uranium was released into the atmosphere and groundwater of the surrounding area, and large amounts of radon and other decay products are thought to have been dispersed into the surrounding air. Household drinking-water sources included reservoirs, such as cisterns, where rainwater was collected from roof gutters. People who lived near the site were potentially exposed to radiologic and nonradiologic toxicants from groundwater, soil contamination, and plant emissions. A medical-surveillance program, the Fernald Resident Medical Monitoring Program (FMMP), was later created to monitor the health of Fernald residents. Pinney and colleagues conducted a study to determine the prevalence of chronic disease in residents who lived near the Fernald facility.

The study cohort was selected from the FMMP, which included adults who lived within 5 miles of the site for 2 years or more during 1952-1984. Former plant workers were excluded, as were 21 FMMP participants who were undergoing chemotherapy at the time of the first FMMP examination. The final study population consisted of 8,464 persons. Medical history (which included current and past medical problems, hospitalizations, surgical and medical procedures, and family history) was ascertained from records of the first FMMP medical examination. Medical conditions were coded according to ICD-9. Additional medical records were consulted to verify some health outcomes. A questionnaire was distributed to gather information on lifestyle risk factors, location of residences and drinking water sources, and other measures of exposure. The questionnaire also included four questions on current medical conditions, such as heart problems, diabetes, cancer, chronic bronchitis, and emphysema. Residential history by location was used to develop a crude exposure metric, which divided participants according to whether they had lived within 2 miles of the facility, lived in the direction of groundwater runoff from the facility, or obtained their drinking water from a well or cistern. The authors conducted interviews with area medical practitioners, examining physicians of the FMMP, local and state health officials, and community residents to obtain information about perceived disease excess.

Selected health outcomes of interest included goiter, other thyroid disease, chronic bronchitis, asthma, emphysema, nephritis, other renal disease, and diabetes mellitus.

Data from the US National Health Interview Study (NHIS) and the Third National Health and Nutrition Examination Survey (NHANES III) were used for comparison. Age- and sex-specific rates were calculated for ICD-9–coded health outcomes by using the FMMP database. Those with more than one code for a specific category were counted once for that category. Separate rates were calculated by using data on the four-question set on the FMMP questionnaire. Both rates were compared with similar rates for white non-Hispanics in the NHANES III and NHIS samples. Sampling weights were applied to account for survey design and nonresponse.

The authors found statistically significant excesses in the FMMP population in renal disease (standardized prevalence ratio [SPR], 215; 99% CI, 186-248), bladder disease (SPR, 132; 99% CI, 111-156), and thyroid diseases (SPR, 155; 99% CI, 133-179). Those outcomes included increases in a few subcategories, such as kidney stones (SPR, 398; 99% CI, 336-468) and chronic nephritis (SPR, 203; 99% CI, 76-435). Residents who obtained their drinking water from a well or cistern had higher urinary microalbumin, hematocrit, and red-cell counts. Residents had significantly fewer cases of asthma (SPR, 85; 99% CI, 73-98), chronic bronchitis (SPR, 19; 99% CI, 14-24), and emphysema (SPR, 61; 99% CI, 41-68) compared with NHIS rates.

Finnish Well-Water Studies

Uranium occurs naturally throughout Earth's crust. Highly concentrated uranium granitoids and granites can become soluble in soft, slightly alkaline bicarbonate groundwater. That is the case in various parts of Finland, where uranium concentrations in drilled wells can reach 12,000 µg/L. The proposed global limit for uranium in drinking water is 2 µg/L. Several animal and occupational studies have documented uranium's toxic effect on the kidneys, specifically on tubular and glomerular function, but few studies have assessed renal toxicity due to uranium exposure through drinking water (Kurttio et al., 2002). Furthermore, animal models show that ingestion of uranium through drinking water increases urinary calcium and phosphate excretion and thus affects bone metabolism (Kurttio et al., 2005).

In six successive studies, Kurttio and colleagues assessed the health effects of naturally occurring uranium in drinking water in 28 municipalities in southern Finland. The municipalities selected had the highest uranium concentrations. The study population was derived from a drinking-water database that contained radioactive measurements for over 5,000 drilled wells (Kurttio et al., 2002). Study subjects were selected through surveys sent to households throughout the area. The first questionnaire, mailed to 798 households, collected details on well-water use, purification methods and equipment, and medical history. A second questionnaire, sent to 436 individuals, limited the cohort to people who resided in households that had no more than two persons and had used a well as the main source of drinking water for at least 1 year. The authors collected details on residential history, well-water consumption and use, education, disease history, smoking history, occupation, and use of medication and herbal products. Of the 436 individuals, 78% consented to participation in the study, and urine and blood samples were collected from them. A third questionnaire was used to obtain additional medical details on bone-fracture history, estrogen use, and previously mentioned items. Uranium exposure was defined with four measures: uranium in drinking water, daily intake of uranium from drinking water, uranium in urine, and estimated cumulative intake from drinking water. Uranium exposure and outcome variables were modeled by using linear regression to establish associations. A summary of the investigations follows.

Kurttio et al., 2002

Researchers investigated possible renal effects by evaluating relevant biomarkers in connection with uranium exposure through drinking water in Finland. The study population (n = 325) consisted of people living in 28 municipalities where uranium concentrations were highest and measurement was frequent. As described above, on the basis of responses from the 798 households, 436 individuals were selected; investigators restricted the population to people older than 15 years who lived in households of no more than two persons that had used a well as the main source of drinking water for at least 1 year. Of the 436 people, 340 (78%) agreed to participate in the study and returned a second questionnaire that detailed SES, smoking history, medical history, drug use, and heavy-metal exposure.

The study was limited to 194 wells in 24 municipalities that served as the main source of drinking water for an average of 13 years (range, 1-34 years). The authors excluded 14 respondents who had diabetes mellitus or used methotrexate or sodium aurothiomalate chronically, pregnant women, and those in households that had efficient purification systems. The study population included 163 women (56% of the group), and the mean age of the group was 52 years (range, 15-82 years). Some 15% of the study population were current smokers; 56% had never smoked cigarettes, cigars, or pipes; and 29% were ex-smokers.

Study participants provided overnight and spot urine and blood samples at least a week after consumption of drilled-well water and water samples for analysis. Investigators also measured blood pressure, height, and body weight. To ensure quality, blinded samples were submitted to different facilities for uranium-isotope measurement, and results were comparable. Four metrics were used in the measurement of uranium exposure: daily intake (of uranium from drinking water), uranium in urine, uranium in drinking water, and estimated cumulative uranium intake from drinking water. A number of metrics were chosen to assess renal function on the basis of previous findings on uranium toxicity. Proximal tubular markers included urinary and serum concentrations of calcium, phosphate, glucose, and beta-2-microglobulin; glomerular function was measured by using creatinine and urinary albumin concentrations. Fractional excretion of calcium and phosphate was used as the primary outcome metric.

The exposure measures were analyzed as both continuous and categorical variables adjusted for age, sex, and body-mass index (BMI) by using general linear-regression models. In determining uranium dose-response relationships, separate analyses were conducted for calcium and phosphate fractional excretion end points for sample points above and below the median. The authors used existing or suggested standards for drinking water to establish cutpoints for uranium in drinking water and approximate values (quintiles) for urinary uranium and daily intake.

Uranium concentration in water ranged from 0.001 to 1,920 μg/L. The median daily intake of uranium from water was 39 μg. Uranium exposure through drinking water was associated only with calcium excretion ($p = 0.03$). Urinary uranium was associated with fractional excretion of calcium for all exposure metrics. The authors also observed an association between fractional phosphate and uranium concentration in urine ($p = 0.03$). There was no association between uranium exposure and measures of glomerular function.

Kurttio et al., 2005

In this study, researchers evaluated biochemical markers of bone turnover in 146 men and 142 women who obtained drinking water from drilled wells in high-uranium areas for an average of 13 years. The study population was a subset of the cohort described in Kurttio et al. (2002). Samples were collected and analyzed as previously described. A third questionnaire was distributed to ascertain additional details on bone fracture, menopause, and physical activity. Details were also collected on estrogen use by 26 women who reported use of oral contraceptives or hormone-replacement therapy. The authors excluded 43 subjects who were less then 25 years old; had diabetes mellitus; reported long-term use of glucocorticoids, thiazide diuretics, methotrexate, or sodium aurothiomalate; were currently pregnant; or had effective water purification equipment. The remaining group consisted of 288 people in 179 households.

Uranium exposure metrics were similar to those used in the 2002 study (daily intake, uranium in urine, uranium in drinking water, and estimated cumulative intake from drinking water); uranium concentration in drinking water served as the primary measure. Indicators of bone formation included serum osteocalcin and amino-terminal propeptide of type 1 procollagen (P1NP) based on an imunoradiometric assay. Bone resorption was assessed according to serum type 1 collagen carboxy-terminal telopeptide (CTx). Urinary calcium and phosphate were also measured. Linear and weighted robust regression methods were used to account for highly influential observations. Separate analyses were conducted for men and women and accounted for age, smoking status, and estrogen use.

Through robust and linear-regression analyses, uranium exposure was shown to be associated with increased CTx in men (uranium in water, $p = 0.05$ and 0.01; daily intake, $p = 0.16$ and 0.02; and cumulative intake, $p = 0.16$ and 0.03, respectively). Uranium concentrations in drinking water appeared to be associated with increased osteocalcin ($p = 0.19$; $p = 0.04$ in linear-regression analysis). Uranium exposure was not related to any biomarkers of bone metabolism in women. P1NP was not associated with uranium exposure.

Kurttio et al., 2006a

In 2003, Kurttio and colleagues continued their assessment of nephrotoxic effects of naturally occurring uranium in well water, focusing on measures associated with renal-cell toxicity and renal tissue damage. The study population was based on a previous study group in which 325 participants were selected in the evaluation of basic renal function (Kurttio et al., 2002). A third questionnaire was sent to the previous study participants; 222 responded, of whom 202 provided samples for a number of tubular and glomerular markers. Details on residential history, daily well-water consumption and use, medical history, use of medication, and smoking history were collected, and only current well-water users were selected. The authors excluded one person who had recently taken methotrexate. The final study population consisted of 95 men and 98 women in 124 households in which drilled wells were the primary source of drinking water for an average of 16 years (range, 5-40 years). The mean age was 56 ± SD 12 years. Some 6% were smokers, and 56% reported never having smoked. The average BMI was 26 ± 4 kg/m^2. Thirty-nine study participants (20%) reported regular use of analgesics 1 year before study enrollment.

Urine, blood, nail, hair, and household water (kitchen-tap) samples were collected from study participants in households that had consumed well water for at least 1 week continuously. Blood pressure, body weight, and height were also measured. As in the previous study, urinary uranium was measured, as were daily intake, daily intake per unit body weight, intake from drinking water, and cumulative intake. A number of biomarkers were measured in urine and serum samples selected as indicators of cell toxicity and renal dysfunction. Cytotoxic measures included N-acetyl-γ-D-glucosaminidase, lactate dehydrogenase, alkaline phosphatase, and γ-glutamyltransferase. Concentrations of α-glutathione-S-transferase, calcium, phosphate, and glucose served as indicators of proximal tubular and glomerular dysfunction. As in the previous study, the authors calculated values for calcium and phosphate fractional excretion, in addition to glucose excretion and creatinine clearance. Linear regression was used to model outcome variables and uranium exposure with adjustment for sex, age (specified as a linear and/or quadratic effect), BMI, smoking, and use of analgesics.

Urinary uranium concentrations were an average of 44% greater than in prior sampling. In general, markers of renal function were within clinical limits. Biomarkers of cytotoxicity, renal proximal tubular function, glomerular function, and other exposure indicators were not statistically significantly associated with urinary uranium concentrations. There were statistically significant associations between cumulative uranium intake and glucose excretion ($p = 0.02$) and between uranium exposure and increased blood pressure (diastolic, $p = 0.01$; systolic, $p = 0.07$).

The studies described above provide useful information on the health effects of uranium exposure from drinking water. The well-water studies have a number of strengths, including the relatively large population, the specificity of the multiple exposure metrics, and the use of individualized clinical evaluation. In addition, the long-term, internal nature of the exposure is analogous to that of veterans exposed to depleted uranium. Despite those strengths, the cross-sectional nature of the exposure assessment limits the ability to ascertain details of chronic health effects. Furthermore, there were large variations in the duration of exposure, and dietary intake was not taken into account in the data-collection process.

Because of some characteristics of granite bedrock, groundwater in Finland has higher amounts of naturally occurring radionuclides (up to 100 or even 1,000 times as high as in other populations). Researchers identified small municipal units in which 90% of residents obtained their water from wells outside the municipal supply. The cohort consisted of those residents of the units who had maintained residence from January 1967 to December 1980 and were born in 1900-1930. The researchers then conducted three nested case-control studies to analyze cancer incidence in correlation with this well-water use.

Auvinen et al., 2002

To assess the effect of natural uranium and other radionuclides in drinking water (internal exposure) on leukemia, researchers identified all leukemia cases in the cohort (well-water drinkers in Finland) who were diagnosed in 1981-1995 (Auvinen et al., 2002). A portion of the noncases in the cohort were chosen for comparison and matched by age and sex to the cases. Well-water use was determined by surveying the subjects and from information from local health officials. Totals of 371 controls and 41 cases were known to have relied extensively on well water before 1981. Samples of drinking water from wells drilled in July-November 1996 were obtained for 274 controls and 35 cases.

Using a modified proportional-hazard model, the authors calculated hazard ratios (HRs) by matching cases at date of diagnosis with controls at risk and assigning controls a weight to represent the entire cohort.

The median uranium concentrations in drinking water were comparable in cases (0.08 Bq/L) and controls (0.06 Bq/L). The median radon concentrations were 80 and 130 Bq/L, respectively. The corresponding median concentrations of radium were 0.01 and 0.03 Bq/L.

The HRs were not statistically significant for water uranium concentration (HR, 0.91; 95% CI, 0.73-1.13 per Bq/L), radon (HR, 0.79; 95% CI; 0.27-2.29 per Bq/L), or radium (HR, 0.80; 95% CI, 0.46-1.39 per Bq/L). Calculations for the groups with the highest exposures also did not show an increased risk of leukemia.

Auvinen et al., 2005

All stomach-cancers in the cohort diagnosed in 1981-1995 were identified by the researchers (Auvinen et al., 2005). Controls were randomly selected from the cohort and matched by age and sex. Some 371 (8.1%) cohort members and 107 (7.2%) stomach cancer cases had used drinking water from drilled wells before 1981. Well-water samples for 274 controls and 88 cases were obtained in July-November 1996.

HRs were calculated by matching cases at date of diagnosis with controls at risk and assigning controls different weights (to reflect more accurately the composition of the entire cohort).

Median total alpha activity in the drinking water was comparable in cases (0.08 Bq/L) and controls (0.05 Bq/L), and uranium concentrations were also comparable (0.07 Bq/L in cases and controls). The HR was not statistically significant for uranium (HR, 0.76; 95% CI, 0.48-1.21 per Bq/L), radium (HR, 0.69; 95% CI, 0.33-1.47 per Bq/L), or radon (HR, 0.68; 95% CI, 0.29-1.59 per 100 Bq/L). The groups with the highest concentrations of radionuclides did not show a statistically significant correlation with stomach cancer.

Kurttio et al., 2006b

From 144,627 persons born between 1900 and 1930 who had lived outside the municipal water supply from 1967 to 1980, researchers identified 884 bladder-cancer and 644 renal-cancer cases, and they selected 4,590 controls randomly from the remaining cohort. Through surveys and local health authorities, they identified 371 (8%) controls, 79 bladder-cancer cases, and 65 renal-cancer cases as having used drilled well water during the specified period. Well-water samples were obtained for 274 (74%) of the controls, 61 (77%) of the bladder-cancer cases, and 51 (78%) of the renal-cancer cases, who made up the study population of this nested case-control investigation. Mean followup time was 19 years.

The effective dose to the kidneys and bladder were calculated on the basis of a consumption rate of 2 L/day; ingestion dose and effective dose were derived from other sources, such as the National Research Council, International Commission on Radiological Protection, and other previously published studies.

Smoking and BMI are known risk factors for renal and bladder cancers, so both were adjusted for in subanalyses (that is, for subjects on whom these data were available). HRs were calculated by matching cases at date of diagnosis with controls at risk and assigning the controls weights to reflect the composition of the entire cohort.

The median effective radiation doses, close to 100 μSv, of all three radionuclides (uranium, radon, and radium), were comparable among bladder-cancer cases, renal-cancer cases, and controls. No statistically significant associations were observed between any of the radionuclides and renal or bladder cancer. When subjects without information on smoking or BMI were excluded from the analysis, the results did not change significantly.

The strengths of this series of nested case-control studies include long-term internal exposure, blinded analysis of the water samples, applicability to the entire cohort (because of weighting of the controls), individual measure of exposure, and data on confounders (for example, smoking history and BMI of some study subjects), allowed assessment of potential bias. Limitations of the study include the lack of data on the amount of water consumed individually, of information on other sources of drinking water (for example, in the workplace), of data on other risk factors (for example, radiation exposures from medical treatment and other sources), and of sufficient statistical power to detect a small risk.

SUMMARY

In assessing the potential health effects of uranium and depleted uranium in humans, the committee examined a number of studies that focused on people who were occupationally exposed to uranium through processing activities, on depleted-uranium exposure in deployed populations, and on uranium exposure through residence and drinking-water use. Each study was carefully reviewed, and the evidence presented was used to elaborate on the key findings and draw conclusions about the relevant malignant and nonmalignant health outcomes in the next chapter.

TABLE 7-1 Studies of Uranium Processors

Study	Design	Population	Person-Years Observed
Cohorts Evaluated in *Gulf War and Health, Volume 1*			
Colorado Plateau (mill workers)			
Wagoner et al., 1964	Cohort	5,370 uranium miners and millers; followup 1950-1962	6,390
Archer et al., 1973	Cohort	662 male uranium-mill workers; followup 1950-1967	
Waxweiler et al., 1983	Cohort	2,002 male uranium millers employed at least 1 day after January 1, 1940, worked at least 1 year in uranium mills, never worked in uranium mining; followup through 1977	43,252
Pinkerton et al., 2004	Cohort	1,484 men who worked in uranium mill at least 1 day after January 1, 1940, worked for at least 1 year in uranium mills, had never worked in a uranium mine; followup through 1998	
Cincinnati, OH (Fernald Feed)			
Boiano et al., 1989	Cross-sectional	146 (70%) of 208 eligible long-term employees at Feed Materials Production Center after releases of uranium oxide from dust collectors in November-December 1984	

Exposure	Outcomes	Adjustments	Comments
Exposure determined by duration and type of employment (miner or miller); for subset of miners, radiation exposure calculated from months of underground experience, estimated dose rate, cumulative dose	Mortality	Age, race	
Occupational exposure to vanadium and long-lived members of uranium radioactive decay series: uranium-238, uranium-234, thorium-230, radium-228, lead-210	Mortality	Age, race, calendar years	
Exposure defined by work history, duration of employment in mills	Mortality	Age, race, sex, calendar period	
Exposure defined by work history, duration of employment	Mortality from NIOSH modified life-table analysis system through 1998	Stratified analyses by duration of employment, time since first employment (latency), year of first employment	Limitations include lack of smoking data, small cohort, limited power to detect moderately increased risk of some outcomes of interest, inability to estimate individual exposures to uranium, silica, vanadium
Self-reported exposure incidents; job history; assessed urinary-uranium data	Lung, renal disease	Smoking	Limitations in exposure based partly on recall; crude and imprecise exposure categories (low, medium, high)

Continued

TABLE 7-1 Continued

Study	Design	Population	Person-Years Observed
Ritz, 1999	Cohort	4,014 white male uranium-processing workers employed in 1951-1989; followup through January 1, 1990	124,177
Oak Ridge, TN			
Polednak and Frome, 1981	Cohort	18,869 white male workers at uranium conversion and enrichment plant, employed in 1943-1947; followup until 1977	
Checkoway et al., 1988	Cohort	6,781 white male employees at nuclear-weapons materials fabrication plant in May 4, 1947-December 31, 1974; followup through 1979	133,535

Exposure	Outcomes	Adjustments	Comments
External radiation exposure derived from film-badge measurements; internal radiation exposure based on combination of individual urine bioassays, environmental area sampling; assessed exposure to uranium, thorium, radium compounds	Cancer mortality from Social Security Administration, National Death Index	Controlled for exposure to trichloroethylene, cutting fluids	Also ran further analyses by radiation dose
Uranium-dust exposure defined by departments and averages where employee worked Range of average concentrations of uranium in air: 25-300 μg/m^3	Mortality	Age, calendar year	
Radiation exposure assessed with film badges, estimates of dose equivalents delivered to lungs obtained from urinalysis measurements, in vivo counting of internally deposited uranium Mean cumulative alpha-radiation dose to lung 8.21 rem, mean cumulative external whole-body penetrating dose from gamma radiation 0.96 rem	Mortality	Age, calendar year	

Continued

TABLE 7-1 Continued

Study	Design	Population	Person-Years Observed
Frome et al., 1990	Cohort	28,008 World War II nuclear-plant workers, white males employed at least 30 days in 1943-December 31, 1947; followup January 1, 1950-January 1, 1979	
Loomis and Wolf, 1996	Cohort (continuation of Checkoway et al., 1988)	6,591 white men, 922 white women, 449 black men, 149 black women, 5 men and women of other racial groups who had worked at plant after 1947; 1,764 white men, 562 white women, 85 black men, 69 black women, 1 man of other racial group who had worked at plant before 1947	
Richardson and Wing, 2006	Cohort; nested case-control	3,864 nuclear-materials fabrication-plant workers employed at least 30 days in May 4, 1947-December 31, 1974; followup through 1990	
Frome et al., 1997	Cohort	106,020 workers at four plants in Oak Ridge, TN, employed at least 30 days in 1943-1985; followup through 1984	

Exposure	Outcomes	Adjustments	Comments
Indexes for radiation by specific job title-department combination	Mortality	SES, duration of employment, facility, birth year, period of followup	
Radiation exposure assessed with film badges, estimates of dose equivalents delivered to lungs obtained from urinalysis measurements, in vivo counting of internally deposited uranium	Mortality	Age, calendar year	
Alpha radiation based on in vivo monitoring, urinalysis results, estimates based on exposure potential, given department in which employed Internal radiation dose: 10-100+ mSv (exposed), <10 (unexposed)	Lung-cancer mortality from National Death Index, Social Security Administration, Health Care Financing Agency, Tennessee Department of Motor Vehicles	Matched for age, year of birth, sex, race, SES, length of employment, employment status	
External radiation exposure based on limited monitoring; internal radiation exposure categorized into three levels: eligible for monitoring but not monitored, eligible for monitoring and monitored, not eligible for monitoring	Mortality	Stratified by length of employment, SES, birth year, age	

Continued

TABLE 7-1 Continued

Study	Design	Population	Person-Years Observed
Four Uranium-Processing Operations			
Dupree et al., 1995	Case-control	787 lung-cancer cases employed at least 183 days at one of four uranium-processing facilities with one control match; followup through 1982	
St. Louis, MO (Mallinckrodt)			
Dupree-Ellis et al., 2000	Cohort	2,514 white male uranium-processing plant workers employed in 1942-1966; followup through 1993	87,757
Portsmouth Uranium Enrichment Facility, Pike County, OH			
Brown and Bloom, 1987	Retrospective cohort	5,773 white male employees who had worked for at least 1 week in September 1954-February 1982	

Exposure	Outcomes	Adjustments	Comments
Alpha radiation from airborne dust containing uranium compounds from employment and occupational monitoring records; gamma radiation from available personal monitoring data	Lung-cancer mortality from death certificates (ICD8A, codes 162.0-163.9)	Matched for race, sex, birth, hire date within 3 years, smoking status, SES (pay code)	Two operations at Y-12 facility in Oak Ridge, TN: one Mallinckrodt Chemical Works Uranium Division, one Feed Materials Production Center in Fernald, OH
Film badges to monitor beta, gamma radiation exposure; 20.8% of workyears had no monitoring results, and exposure was estimated with algorithm Median cumulative whole-body exposure 15.3 mSv	Mortality from National Death Index, Social Security Administration, Pension Benefit Information through 1993	Age, calendar period	Cohort overlaps with Dupree et al., 1995
Urinary-uranium data (collected since 1954) used to identify departments with potential exposure; departments ranked by relative degree of potential exposure Other data included continuous air sampling, personal sampling of external radiation with film badges, dosimeters, in vivo counting	Mortality	Age, calendar time	Short followup (maximum 17 years) limited ability to assess cancer outcomes Study included in *Gulf War and Health, Volume 1*, but additional cancer outcomes included in current report

Continued

TABLE 7-1 Continued

Study	Design	Population	Person-Years Observed
Polk County, FL (phosphate-fertilizer production workers)			
Stayner et al., 1985	Retrospective cohort	3,199 ever employed at phosphate-fertilizer production facility in Polk County, FL; subjects followed from first date of hire to December 31, 1977, or date of death	
Connecticut (nuclear-fuels fabrication plant)			
Hadjimichael et al., 1983	Retrospective cohort	4,106 employees of plant who had ever worked for more than 6 months in 1956-1978	
New Cohorts (Not Evaluated in *Gulf War and Health, Volume 1*)			
UK (Springfields)			
McGeoghegan and Binks, 2000b	Cohort	19,454 current and former British Nuclear Fuels workers ever employed before 1996; followup 1946-December 31, 1995	479,146

Exposure	Outcomes	Adjustments	Comments
	Mortality	Age, sex, race, calendar year	Lack of exposure information Study included in *Gulf War and Health, Volume 1*, but additional cancer outcomes included in current report
Exposure groups based on film badges worn by employees, job classification, consultation with industrial hygiene, safety personnel, supervisors, employees who had been at plant since operation began Job experience obtained from personnel records	Mortality from Social Security Administration records, Connecticut mortality records; cancer incidence from Connecticut Tumor Registry	Age, sex, calendar year, cause of death or cancer site	Potential exposure overlap between groups due to multiple exposures; did not quantify degree of exposure for individuals or groups Study included in *Gulf War and Health, Volume 1*, but additional cancer outcomes included in current report
Radiation dosimetry collected for compliance with statutory radiation-protection guidelines; used film badges and recorded mSv; included doses acquired from other sites for workers transferred into plant Mean individual cumulative external whole-body dose: 22.8 mSv (analysis A), 20.5 mSv (analysis B)	Mortality, cancer morbidity from 1971-December 31, 1995	Age, sex, calendar year, industrial status (industrial or nonindustrial), worker status, year of joining, time from first exposure, length of service	Additional analyses by dose, lag time

Continued

TABLE 7-1 Continued

Study	Design	Population	Person-Years Observed
UK (Chapelcross)			
McGeoghegan and Binks, 2001	Cohort	2,628 nuclear-fuels workers ever employed in 1955-1995; followup 1955-December 31, 1995	
UK (Capenhurst)			
McGeoghegan and Binks, 2000a	Cohort	12,543 workers at British Nuclear Fuels plant ever employed before 1996; followup 1946-December 31, 1995	334,473
Rocketdyne (Atomics International)			
Ritz et al., 2000	Retrospective cohort	2,297 male (2,218) and female (79) employees enrolled in company's health-physics radiation-monitoring program in January 1, 1950-December 31, 1993; followup through 1994	58,837

Exposure	Outcomes	Adjustments	Comments
Radiation dosimetry collected for compliance with statutory radiation-protection guidelines; used film badges and recorded mSv; included doses acquired from other sites for workers transferred into plant Mean cumulative external whole-body dose: 83.6 mSv	Mortality, cancer morbidity through December 31, 1995	Age, sex, calendar year, industrial status (industrial or nonindustrial), workers status, year of joining, time from first exposure, length of service	Additional analyses by dose, lag time
Radiation dosimetry collected for compliance with statutory radiation-protection guidelines; used film badges and recorded mSv; included doses acquired from other sites for workers transferred into plant Mean cumulative external whole-body dose: 9.85 mSv	Mortality, cancer morbidity 1971-December 31, 1995	Age, sex, calendar year, industrial status (industrial or nonindustrial), workers status, year of joining, time from first exposure, length of service	Additional analyses by dose, lag time
Doses estimated by urine or feces bioassay, in vivo whole-body counts, lung counts Estimated internal cumulative dose to lung of each employee was calculated Mean lung dose: 2.1 mSv	Mortality	Pay categories used as proxy for SES, smoking status Company did not collect information on race before 1972, although 96% of deceased workers were white	

Continued

TABLE 7-1 Continued

Study	Design	Population	Person-Years Observed
Boice et al., 2006a	Retrospective cohort	5,801 radiation workers monitored for radiation and employed on or after January 1, 1948, for at least 6 months; person-years of followup began 6 months after date of first radiation monitoring or July 1, 1948, and stopped at date of death, December 31, 1999, age of 95 years, or date lost to followup	161,605
Savannah River Plant			
Cragle et al., 1988	Retrospective cohort	9,860 white male employees who worked at plant in 1952-1981 and were hired before 1975; followup through 1980	232,061
Atomic Weapons Establishment, UK			
Beral et al., 1988	Retrospective cohort	22,552 employees who worked at establishment in January 1, 1951-December 31, 1982; followup through 1982	
Egyptian Processors			
Shawky et al., 2002	Cross-sectional	86 processors at two sites in Egypt working in three locations, 13 of whom also participated in urinary-uranium analysis	

Exposure	Outcomes	Adjustments	Comments
Annual doses of internally deposited radionuclides estimated by positive bioassay data, in vivo lung counts, incident reports	Mortality	Race, age, calendar year, sex	Low doses, small study, incomplete smoking histories, imperfect categorization of pay type, dosimetry sources not evaluated
Internal monitoring of 2,232 of 5,801 workers; measurements included radionuclides in urine, supplemented with fecal measurements, lung counts			
Exposure defined by work history, duration of employment (average exposure, 13 years)	Mortality	Stratified by duration of employment, date of first employment, hourly vs employment type (hourly vs salaried)	No exposure assessment, lack of generalizability
Exposure measured with dosimeters worn externally	Mortality	Age, sex, calendar period, social class	
Average exposures per radiation worker, 7.8 mSv (whole-body exposure), 14.4 mSv (surface exposure)			
Concentration of uranium in air: 22.6 $\times 10^{-7} - 11.1 \times 10^{-5}$ Bq/cm^3	Clinical measurements of hematologic, renal function		Tested air uranium concentration
Exposure: 1-80 μSv/h			

TABLE 7-2 Studies of Depleted-Uranium–Exposed Persons

Study	Design	Population	Exposure
Baltimore VA Medical Center Study			
McDiarmid et al., 2000	Case series	29 exposed GW veterans exposed to DU during friendly-fire incidents in February 1991, 38 DU-nonexposed veterans; examined in March-June 1997, 7 years after first exposure	Exposure to DU by means of friendly fire; may have inhaled or ingested airborne DU particles, and/or experienced wound contamination by DU; assessed urine and semen uranium concentration Veterans with DU fragments: 0.01-30.7 μg/g creatinine vs nonexposed: 0.01-0.05 μg/g creatinine
McDiarmid et al., 2001	Case series	50 exposed GW veterans divided into low-uranium, high-uranium groups; examined in March-July 1999, 8 years after first exposure	Exposure to DU by means of friendly fire; may have inhaled or ingested airborne DU particles, and/or experienced wound contamination by DU; assessed urine uranium concentration Veterans with DU fragments: 0.018-39.1 μg/g creatinine vs DU-exposed veterans without fragments: 0.002-0.231 μg/g creatinine
McDiarmid et al., 2002	Case series	29 exposed GW veterans, 38 nonexposed GW veterans, 30 newly identified exposed; examined in spring 1999	Exposure to DU by friendly fire; may have inhaled or ingested airborne DU particles, experienced wound contamination by DU; assessed urinary uranium
McDiarmid et al., 2004	Case series	39 GW veterans exposed to DU during friendly-fire incidents in February 1991; examined in April-July; followup in 1994-2001	Exposure to DU by friendly fire; may have inhaled, ingested airborne DU particles, experienced wound contamination by DU; assessed urinary uranium Low-uranium group, <0.1 μg/g of creatinine; high-uranium group, ≥0.1 μg/g of creatinine

Outcomes	Adjustments	Comments
Hematologic, renal function; neurocognitive, psychiatric measures; genotoxicity measures; concentrations of follicle stimulating hormone, prolactin, testosterone; semen characteristics (volume, concentration, morphology, motility)	Stratification at median into low-, high-result groups	Small sample
Hematologic, renal, neurocognitive, genotoxic outcome measures; injuries; concentrations of thyroid-stimulating hormone, free thyroxine; reproductive neuroendocrine indicators, semen characteristics reported in McDiarmid et al., 2000	Race, education, age, marital status, military rank, intelligence (WRAT-3), depression (BDI), smoking status, use of prescription psychotropic, antidepressant drugs, recent X-ray exposure	Small sample, no true comparison group
Urinary uranium determinations; clinical laboratory values; psychiatric, neurocognitive assessment		Small sample
Hematologic, renal function; immunologic measures; genetoxicity; neurocognitive, psychiatric assessment; reproductive characteristics reported in McDiarmid et al., 2001	Age, smoking, exposure to genetic toxicants, cloning efficiency	No comparison group, small sample

Continued

TABLE 7-2 Continued

Study	Design	Population	Exposure
McDiarmid et al., 2006	Case series	32 GW veterans exposed to DU during friendly fire; examined in April-July 2003	Exposure to DU by friendly fire; may have inhaled, ingested airborne DU particles, experienced wound contamination by DU; assessed urinary uranium
			Low-uranium group, <0.1 μg/g of creatinine; high-uranium group, ≥0.1 μg/g of creatinine
McDiarmid et al., 2007	Case series	34 GW veterans exposed to DU during friendly-fire incidents in 1991; examined in April-June 2005	Exposure to DU by friendly fire; may have inhaled, ingested airborne DU particles, experienced wound contamination by DU; assessed urinary uranium; both current and cumulative exposure measures reported
			Low-uranium group, <0.1 μg/g of creatinine; high-uranium group, ≥0.1 μg/g of creatinine
UK Gulf War Veterans			
Macfarlane et al., 2003	Cohort	51,721 UK GW veterans, 50,755 nondeployed UK service personnel; followup April 1, 1991-July 31, 2002	Deployment to GW; self-reported exposure to DU
Balkans Cohorts			
Gustavsson et al., 2004	Cohort	9,188 Swedish military personnel deployed to UN missions in Balkans in 1989-1999; followed up through December 31, 1999; 39,816 person-years	Deployment to UN missions in Balkans
Nuccetelli et al., 2005	Summary of data presented in Italian Defence Ministry study published in Italian in 2002	40,000 Italian soldiers deployed to Balkans at least once in 1995-2001 (followup time not reported)	Deployment to Balkans

Outcomes	Adjustments	Comments
Hematologic characteristics; renal function; neurocognitive measures; genotoxicity; reproductive characteristics reported in McDiarmid et al., 2001	Age, intelligence, emotional status, smoking, exposure to genetic toxicants, cloning efficiency	Small sample, no true comparison group
Hematologic, renal function; neurocognitive, psychiatric measures; genotoxicity measures; reproductive characteristics reported in McDiarmid et al., 2001	Age, IQ, depression, cloning efficiency	Small sample, no true comparison group
Cancer	DU analysis adjusted for smoking, alcohol consumption; matched for age, sex, rank, service, level of fitness	Latency of many cancers is beyond time of study
Cancer	Sex, age, period	Short followup period for cancer
Cancer		

Continued

TABLE 7-2 Continued

Study	Design	Population	Exposure
Storm et al., 2006	Population-based retrospective cohort	Danish military deployed to Balkans (13,552 men, 460 women); followup from January 2002 to December 2002	Deployment to Balkans
Sumanovic-Glamuzina et al., 2003	Pre-post comparison	All liveborn and stillborn neonates in Maternity Ward of Mostar University Hospital in western Herzegovina, part of Bosnia and Herzegovina, immediately (1995) and 5 years after (2000) 1991-1995 military activities	Living in western Herzegovina after military activities

Outcomes	Adjustments	Comments
Cancer	Age-, sex-, period-specific SIRs	Few cases, wide CIs, young cohort
Major congenital malformations		Not known whether DU was used in region

TABLE 7-3 Studies of Environmental Exposure to Uranium

Reference	Design	Population	Exposure
Residential Studies			
Bithell and Draper, 1999	Modeling	Those living within 6 km of Air Force base	Environmental uranium (for example, in leaves)
Boice et al., 2003a	Cross-sectional	16,722 people living in 8 municipalities in PA; cancer-mortality records from Pennsylvania Cancer Registry for 1993-1997; reference, PA or national SEER registry	Residential proximity to Apollo, Parks nuclear facilities in PA
Boice et al., 2003c	Ecologic mortality survey	Mortality data from NCI, Texas Department of Health for 1950-2001 Cases, Karnes County (site of uranium mining); controls: 4 counties matched on various characteristics; reference, US general population	Residential proximity to uranium-processing site
Boice et al., 2003b	Mortality survey	Cases: 3 comparison counties matched on demographics; controls, 2 counties in PA from NCHS for 1950-1995; reference, US general population	Residential proximity to Apollo, Parks nuclear facilities
Pinney et al., 2003	Cohort	8,464 people from FMMP; comparison rates, NHIS, NHANES	Residential proximity (less than 2 miles) to Fernald uranium-processing plant in direction of groundwater runoff or possible well or cistern contamination in January 1952-December 1984

Outcomes	Adjustments	Comments
Childhood leukemia		Authors modeled uranium exposure around US Air Force base and compared it with distribution of childhood leukemia in same area; no correlation found between two plots
Cancer incidence	Age, sex, calendar year	No adjustment for diet, smoking, other cancer risk factors; no determination of length of residence (and hence exposure)
Cancer mortality	Control counties chosen on basis of matched demographics	No adjustment for diet, smoking, other cancer risk factors; no determination of length of residence (and hence exposure)
Cancer mortality	Matched on demographics	No adjustment for diet, smoking, other cancer risk factors; no determination of length of residence (and hence exposure)
Goiter, other thyroid disease, chronic bronchitis, asthma, emphysema, nephritis, other renal disease, diabetes mellitus	Age, sex	Study questionnaires not directly comparable; FMMP is self-selected volunteer group

Continued

TABLE 7-3 Continued

Reference	Design	Population	Exposure
Finland Well-Water Studies			
Kurttio et al., 2002	Cross-sectional	325 people in Finland who obtain drinking water from drilled wells used an average of 13 years	Median drinking-water uranium concentration, 28 μg/L (interquartile range, 6-135 μg/L; maximum, 1,920 μg/L) Median urinary uranium concentration, 13 ng/mmol of creatinine (range, 2-75 ng/mmol) Median daily uranium intake, 39 μg (range, 7-224 μg)
Kurttio et al., 2005 (same population as Kurttio et al., 2002)	Cross-sectional	146 men, 142 women in southern Finland who obtain drinking water from drilled wells used an average of 13 years	Median drinking-water uranium concentration, 27 μg/L (interquartile range, 6-116 μg/L) Median daily uranium intake, 36 μg (range, 7-207 μg) Median cumulative intake, 0.12 g (range, 0.02-0.66 g)
Kurttio et al., 2006a (same population as Kurttio et al., 2002)	Cross-sectional	95 men, 98 women in Finland who obtain drinking water from drilled wells used an average of 16 years	Median drinking-water uranium concentration, 25 μg/L (interquartile range, 5-148 μg/L; maximum, 1,500 μg/L)
Auvinen et al., 2002	Nested case-control	35 leukemia cases, 274 stratified randomly sampled people from subcohort who obtained well water before 1981	Well-water samples collected blind in July-November 1996 Median activity uranium concentration for leukemia cases, 0.08 Bq/L; for reference group, 0.06 Bq/L

Outcomes	Adjustments	Comments
Renal function (based on urinary, serum concentrations of calcium, phosphate, glucose, albumin, creatinine, beta-2-microglobulin as biomarkers)	Uranium exposure adjusted for age, sex, BMI	Uranium exposure measured by daily intake, uranium in urine, uranium in drinking water, cumulative intake from drinking water
Indicators of bone formation: serum osteocalcin, amino-terminal of type 1 procollagen (P1NP); indicators of bone resorption: serum type 1 collagen carboxy-terminal telopeptide (CTx), urinary calcium, urinary phosphate	Age, smoking, estrogen use (women)	Used two types of regression analysis (linear regression, weighted robust regression) to account for highly influential observations
Renal-cell toxicity, renal dysfunction (based on concentrations of various enzymes, creatinine, calcium, phosphate, glucose as indicators) Exposure assessment: uranium concentration in drinking water, hair, nails, urine	Sex, age (linear-quadratic), BMI, smoking, use of analgesics	Reference group is general population
Association with uranium, radon, radium	Age, sex	No dose-response assessment; no adjustments for other risk factors

Continued

TABLE 7-3 Continued

Reference	Design	Population	Exposure
Auvinen et al., 2005	Nested case-control	88 stomach-cancer cases, 274 stratified randomly sampled people from subcohort who obtained water from drilled wells before 1981	Well-water samples collected blind in July-November 1996 Median activity concentration for both cases and reference group, 130 Bq/L
Kurttio et al., 2006b	Nested case-control	Cases, 61 bladder-cancer cases, 51 renal-cancer cases diagnosed in 1981-1995; controls, 274 randomly sampled people stratified by sex, age	Drilled well water outside municipal water supply obtained in 1967-1980 Uranium concentrations: bladder cancer, 0.08 Bq/L; renal cancer,0.07 Bq/L; reference, 0.06 Bq/L

Outcomes	Adjustments	Comments
Association with uranium, radon, radium	Age, sex	No dose-response assessment; no adjustments for other risk factors
Association with radon, radium, uranium exposure	Bladder cancer: age at followup, sex, smoking status Renal cancer: age, sex, smoking, BMI	Exposures measured only up to 10 years before diagnosis to account for cancer latency

REFERENCES

Archer, V. E., J. K. Wagoner, and F. E. Lundin, Jr. 1973. Cancer mortality among uranium mill workers. *Journal of Occupational Medicine* 15(1):11-14.

Auvinen, A., P. Kurttio, J. Pekkanen, E. Pukkala, T. Ilus, and L. Salonen. 2002. Uranium and other natural radionuclides in drinking water and risk of leukemia: A case-cohort study in Finland. *Cancer Causes and Control* 13(9):825-829.

Auvinen, A., L. Salonen, J. Pekkanen, E. Pukkala, T. Ilus, and P. Kurttio. 2005. Radon and other natural radionuclides in drinking water and risk of stomach cancer: A case-cohort study in Finland. *International Journal of Cancer* 114(1):109-113.

Beral, V., P. Fraser, L. Carpenter, M. Booth, A. Brown, and G. Rose. 1988. Mortality of employees of the Atomic Weapons Establishment, 1951-82. *British Medical Journal* 297(6651):757-770.

Bithell, J. F., and G. J. Draper. 1999. Uranium-235 and childhood leukaemia around Greenham Common airfield. *Journal of Radiological Protection* 19(3):253-259.

Boiano, J. M., C. E. Moss, and G. A. Burr. 1989. *Health hazard evaluation report HETA 83-144-2001, feed materials production center*. Fernald, OH: Westinghouse Materials Company of Ohio.

Boice, J. D., Jr., W. L. Bigbee, M. T. Mumma, and W. J. Blot. 2003a. Cancer incidence in municipalities near two former nuclear materials processing facilities in Pennsylvania. *Health Physics* 85(6):678-690.

———. 2003b. Cancer mortality in counties near two former nuclear materials processing facilities in Pennsylvania, 1950-1995. *Health Physics* 85(6):691-700.

Boice, J. D., Jr., M. Mumma, S. Schweitzer, and W. J. Blot. 2003c. Cancer mortality in a Texas county with prior uranium mining and milling activities, 1950-2001. *Journal of Radiological Protection* 23(3):247-262.

Boice, J. D., S. S. Cohen, M. T. Mumma, E. Dupree Ellis, K. F. Eckerman, R. W. Leggett, B. B. Boecker, A. B. Brill, and B. E. Henderson. 2006a. Mortality among radiation workers at Rocketdyne (Atomics International), 1948-1999. *Radiation Research* 166(1 Pt 1):98-115.

Boice, J. D., Jr., R. W. Leggett, E. D. Ellis, P. W. Wallace, M. Mumma, S. S. Cohen, A. B. Brill, B. Chadda, B. B. Boecker, R. C. Yoder, and K. F. Eckerman. 2006b. A comprehensive dose reconstruction methodology for former Rocketdyne/Atomics International radiation workers. *Health Physics* 90(5):409-430.

Brown, D. P., and T. Bloom. 1987. *Mortality among uranium enrichment workers*. Cincinnati, OH: National Institute of Occupational Safety and Health.

Checkoway, H., N. Pearce, D. J. Crawford-Brown, and D. L. Cragle. 1988. Radiation doses and cause-specific mortality among workers at a nuclear materials fabrication plant. *American Journal of Epidemiology* 127(2):255-266.

Cookfair, D. L., W. L. Beck, C. Shy, C. C. Lushbough, and C. L. Sowder. 1983. Lung cancer among workers at a uranium processing plant. Presented at Epidemiology Applied to Health Physics. In *Proceedings of the Health Physics Society*. Pp. 398-406.

Cragle, D. L., R. W. McLain, J. R. Qualters, J. L. S. Hickey, G. S. Wilinson, W. G. Tankersley, and C. C. Lushbaugh. 1988. Mortality among workers at a nuclear fuels production facility. *American Journal of Industrial Medicine* 14:379-401.

Dupree, E. A., D. L. Cragle, R. W. McLain, D. J. Crawford-Brown, and M. J. Teta. 1987. Mortality among workers at a uranium processing facility, the Linde Air Products company ceramics plant, 1943-1949. *Scandinavian Journal of Work, Environment and Health* 13(2):100-107.

Dupree, E. A., J. P. Watkins, J. N. Ingle, P. W. Wallace, C. M. West, and W. G. Tankersley. 1995. Uranium dust exposure and lung cancer risk in four uranium processing operations. *Epidemiology* 6(4):370-375.

Dupree-Ellis, E., J. Watkins, J. N. Ingle, and J. Phillips. 2000. External radiation exposure and mortality in a cohort of uranium processing workers. *American Journal of Epidemiology* 152(1):91-95.

Frome, E. L., D. L. Cragle, and R. W. McLain. 1990. Poisson regression analysis of the mortality among a cohort of World War II nuclear industry workers. *Radiation Research* 123(2):138-152.

Frome, E. L., D. L. Cragle, J. P. Watkins, S. Wing, C. M. Shy, W. G. Tankersley, and C. M. West. 1997. A mortality study of employees of the nuclear industry in Oak Ridge, Tennessee. *Radiation Research* 148(1):64-80.

Gustavsson, P., M. Talback, A. Lundin, B. Lagercrantz, P. E. Gyllestad, and L. Fornell. 2004. Incidence of cancer among Swedish military and civil personnel involved in UN missions in the Balkans 1989-99. *Occupational and Environmental Medicine* 61(2):171-173.

Hadjimichael, O. C., A. M. Ostfeld, D. A. D'Atri, and R. E. Brubaker. 1983. Mortality and cancer incidence experience of employees in a nuclear fuels fabrication plant. *Journal of Occupational Medicine* 25(1):48-61.

IOM (Institute of Medicine). 2000. *Gulf War and health, volume 1: Depleted uranium, sarin, pyridostigmine bromide, vaccines*. Washington, DC: National Academy Press.

Kurttio, P., A. Auvinen, L. Salonen, H. Saha, J. Pekkanen, I. Makelainen, S. B. Vaisanen, I. M. Penttila, and H. Komulainen. 2002. Renal effects of uranium in drinking water. *Environmental Health Perspectives* 110(4):337-342.

Kurttio, P., H. Komulainen, A. Leino, L. Salonen, A. Auvinen, and H. Saha. 2005. Bone as a possible target of chemical toxicity of natural uranium in drinking water. *Environmental Health Perspectives* 113(1):68-72.

Kurttio, P., A. Harmoinen, H. Saha, L. Salonen, Z. Karpas, H. Komulainen, and A. Auvinen. 2006a. Kidney toxicity of ingested uranium from drinking water. *American Journal of Kidney Disease* 47(6):972-982.

Kurttio, P., L. Salonen, T. Ilus, J. Pekkanen, E. Pukkala, and A. Auvinen. 2006b. Well water radioactivity and risk of cancers of the urinary organs. *Environmental Research* 102(3):333-338.

Loomis, D. P., and S. H. Wolf. 1996. Mortality of workers at a nuclear materials production plant at Oak Ridge, Tennessee, 1947-1990. *American Journal of Industrial Medicine* 29(2):131-141.

Macfarlane, G. J., E. Thomas, and N. Cherry. 2000. Mortality among UK Gulf War veterans. *Lancet* 356(9223):17-21.

Macfarlane, G. J., A. M. Biggs, N. Maconochie, M. Hotopf, P. Doyle, and M. Lunt. 2003. Incidence of cancer among UK Gulf War veterans: Cohort study. *British Medical Journal* 327(7428):1373.

McDiarmid, M. 2007 (unpublished). *Depleted uranium (DU) follow-up program update. Presentation to the IOM Committee on Gulf War and Health: Depleted Uranium*. College Park, MD: University of Maryland.

McDiarmid, M. A., J. P. Keogh, F. J. Hooper, K. McPhaul, K. Squibb, R. Kane, R. DiPino, M. Kabat, B. Kaup, L. Anderson, D. Hoover, L. Brown, M. Hamilton, D. Jacobson-Kram, B. Burrows, and M. Walsh. 2000. Health effects of depleted uranium on exposed Gulf War veterans. *Environmental Research* 82(2):168-180.

McDiarmid, M. A., K. Squibb, S. Engelhardt, M. Oliver, P. Gucer, P. D. Wilson, R. Kane, M. Kabat, B. Kaup, L. Anderson, D. Hoover, L. Brown, and D. Jacobson-Kram. 2001. Surveillance of depleted uranium exposed Gulf War veterans: Health effects observed in an enlarged "friendly fire" Cohort. *Journal of Occupational and Environmental Medicine* 43(12):991-1000.

McDiarmid, M. A., F. J. Hooper, K. Squibb, K. McPhaul, S. M. Engelhardt, R. Kane, R. DiPino, and M. Kabat. 2002. Health effects and biological monitoring results of Gulf War veterans exposed to depleted uranium. *Military Medicine* 167(2 Suppl):123-124.

McDiarmid, M. A., S. Engelhardt, M. Oliver, P. Gucer, P. D. Wilson, R. Kane, M. Kabat, B. Kaup, L. Anderson, D. Hoover, L. Brown, B. Handwerger, R. J. Albertini, D. Jacobson-Kram, C. D. Thorne, and K. S. Squibb. 2004. Health effects of depleted uranium on exposed Gulf War veterans: A 10-year follow-up. *Journal of Toxicology and Environmental Health A* 67(4):277-296.

McDiarmid, M. A., S. M. Engelhardt, M. Oliver, P. Gucer, P. D. Wilson, R. Kane, M. Kabat, B. Kaup, L. Anderson, D. Hoover, L. Brown, R. J. Albertini, R. Gudi, D. Jacobson-Kram, C. D. Thorne, and K. S. Squibb. 2006. Biological monitoring and surveillance results of Gulf War I veterans exposed to depleted uranium. *International Archives of Occupational and Environmental Health* 79(1):11-21.

McDiarmid, M. A., S. M. Engelhardt, M. Oliver, P. Gucer, P. D. Wilson, R. Kane, A. Cernich, B. Kaup, L. Anderson, D. Hoover, L. Brown, R. Albertini, R. Gudi, D. Jacobson-Kram, and K. S. Squibb. 2007. Health surveillance of Gulf War I veterans exposed to depleted uranium: Updating the cohort. *Health Physics* 93(1):60-73.

McGeoghegan, D., and K. Binks. 2000a. The mortality and cancer morbidity experience of workers at the Capenhurst uranium enrichment facility 1946-95. *Journal of Radiological Protection* 20(4):381-401.

———. 2000b. The mortality and cancer morbidity experience of workers at the Springfields uranium production facility, 1946-95. *Journal of Radiological Protection* 20(2):111-137.

———. 2001. The mortality and cancer morbidity experience of employees at the Chapelcross plant of British Nuclear Fuels PLC, 1955-95. *Journal of Radiological Protection* 21(3):221-250.

Nuccetelli, C., M. Grandolfo, and S. Risica. 2005. Depleted uranium: Possible health effects and experimental issues. *Microchemical Journal* 79(1-2):331-335.

Pinkerton, L. E., T. F. Bloom, M. J. Hein, and E. M. Ward. 2004. Mortality among a cohort of uranium mill workers: An update. *Occupational and Environmental Medicine* 61(1):57-64.

Pinney, S. M., R. W. Freyberg, G. E. Levine, D. E. Brannen, L. S. Mark, J. M. Nasuta, C. D. Tebbe, J. M. Buckholz, and R. Wones. 2003. Health effects in community residents near a uranium plant at Fernald, Ohio, USA. *International Journal of Occupational Medicine and Environmental Health* 16(2):139-153.

Polednak, A. P., and E. L. Frome. 1981. Mortality among men employed between 1943 and 1947 at a uranium-processing plant. *Journal of Occupational Medicine* 23(3):169-178.

Richardson, D. B., and S. Wing. 2006. Lung cancer mortality among workers at a nuclear materials fabrication plant. *American Journal of Industrial Medicine* 49(2):102-111.

Ritz, B. 1999. Radiation exposure and cancer mortality in uranium processing workers. *Epidemiology* 10(5):531-538.

Ritz, B., H. Morgenstern, D. Crawford-Brown, and B. Young. 2000. The effects of internal radiation exposure on cancer mortality in nuclear workers at Rocketdyne/Atomics International. *Environmental Health Perspectives* 108(8):743-751.

Shawky, S., H. A. Amer, M. I. Hussein, Z. el-Mahdy, and M. Mustafa. 2002. Uranium bioassay and radioactive dust measurements at some uranium processing sites in Egypt—health effects. *Journal of Environmental Monitoring* 4(4):588-591.

Stayner, L. T., T. Meinhardt, R. Lemen, D. Bayliss, R. Herrick, G. R. Reeve, A. B. Smith, and W. Halperin. 1985. A retrospective cohort mortality study of a phosphate fertilizer production facility. *Archives of Environmental Health* 40(13):133-138.

Storm, H. H., H. O. Jorgensen, A. M. Kejs, and G. Engholm. 2006. Depleted uranium and cancer in Danish Balkan veterans deployed 1992-2001. *European Journal of Cancer* 42(14):2355-2358.

Sumanovic-Glamuzina, D., V. Saraga-Karacic, Z. Roncevic, A. Milanov, T. Bozic, and M. Boranic. 2003. Incidence of major congenital malformations in a region of Bosnia and Herzegovina allegedly polluted with depleted uranium. *Croatian Medical Journal* 44(5):579-584.

Wagoner, J. K., V. E. Archer, B. E. Carroll, D. A. Holaday, and P. A. Lawrence. 1964. Cancer mortality patterns among U.S. uranium miners and millers, 1950 through 1962. *Journal of the National Cancer Institute* 32(4):787-801.

Waxweiler, R. J., V. E. Archer, R. J. Roscoe, A. Watanabe, and M. J. Thun. 1983. Mortality patterns among a retrospective cohort of uranium mill workers. In *Epidemiology Applied to Health Physics,* Proceedings of the Sixteenth Midyear Topical Meeting of the Health Physics Society, Albuquerque, New Mexico, January 9-13:428-435.

8

Conclusions

In this chapter, the committee further evaluates the peer-reviewed published literature to draw conclusions about the long-term human health outcomes associated with exposure to natural uranium (as occurred in uranium-processing mills and other facilities and in residences) or depleted uranium (as occurred in the Gulf War). The discussion is organized according to cancer (or malignant) and noncancer (or nonmalignant) health outcome. Tables included at the end of this chapter contain results from the studies on which the committee bases its conclusions.

The traditional 5% level of statistical significance is used in describing the committee's conclusions regarding associations. Associations that did not reach the 5% level of statistical significance are described below as nonsignificant.

CANCER OUTCOMES

This section presents the strength of associations between exposure to natural or depleted uranium and particular cancer outcomes. It draws on the information from the many studies that were described in Chapter 7 and on *Gulf War and Health, Volume 1: Depleted Uranium, Pyridostigmine Bromide, Sarin, Vaccines* (IOM, 2000; hereafter referred to as *Volume 1*). The committee focused on the following sites: leukemias, lymphomas, and cancers of the lung, bone, kidney, bladder, stomach, central nervous system, prostate, and testis.

Most of the studies examined cancer mortality, but several studies of UK Gulf War veterans, Balkans veterans, and the Finnish drinking-water cohort also investigated cancer incidence. Because several cancers of interest are associated with a generally good chance of survival, cancer incidence (ascertainable from cancer-registration programs) is a better indicator of cancer risk than cancer-related mortality.

Results of cancer studies conducted in animal models are inconsistent (see Chapter 3). Several studies reported positive findings with respect to the development of a variety of cancers (including lung and renal cancers, leukemia, and sarcoma) in animals exposed by inhalation of uranium-ore dust or uranium dioxide, intratracheal injection of ^{235}U (as tetravalent or hexavalent uranium), or implantation of depleted-uranium pellets (Leach et al., 1973; Filippova et al., 1978; Mitchel et al., 1999; Hahn et al., 2002; Miller et al., 2005). However, other studies reported no increase in tumor development in animals exposed by inhalation of uranium-ore dust or ingestion of uranium (Maynard and Hodge, 1949; Cross et al., 1981; ATSDR, 1999).

Lung Cancer

Twenty-three studies of uranium-processing workers examined the association between exposure to uranium and lung cancer, as did three studies of military populations and three studies of residents (see Table 8-1). Four of the uranium-processing studies reported statistically significantly increased standardized mortality ratios (SMR) (that is, above 100). All four of those studies involved the same cohort of Oak Ridge, Tennessee, and all included employees of the Y-12 plant (see Table 8-2). The specific study populations overlapped, but each study took a different approach and examined a different timeframe. The most recent study of the cohort, by Richardson and Wing (2006), did not demonstrate a statistically significant increase in lung-cancer mortality in any dose stratum. However, when assessing the dose-response relationship with a 5-year lag assumption, they found a dose-response trend between external exposure and lung-cancer mortality (due largely to a small number of excess deaths among those who accumulated an external dose of 50 mSv or more) but did not find a similar trend for internal exposure. Analyses of the joint effects of external and internal exposures found that compared to the referent group (defined as less than 10 mSv external and internal dose), the rate ratio estimates were increased for each group defined by higher cumulative concentrations of internal and/or external dose; however, the results were not statistically significant and a dose-response trend was not observed. One major limitation of the uranium-processing worker studies is the lack of control for smoking, a major risk factor for lung cancer.

Contrary to the Y-12 cohort finding, a UK study of processors found significant reductions in both mortality from lung cancer (SMR, 85; $p < 0.05$) and incidence of lung cancer (standardized incidence ratio [SIR], 75; $p < 0.001$) but is limited by having only external-exposure data (McGeoghegan and Binks, 2000b). Beral et al. (1988) also reported a significant deficit in lung-cancer mortality (SMR, 64; $p < 0.01$) in employees of UK atomic-weapons research establishments with radiation records but found a significant positive association between cumulative exposure and lung-cancer mortality in a test for trend. One study of residents living near former nuclear-material processing plants found a significant reduction in risk of lung-cancer death (relative risk [RR], 0.95; 95% confidence interval [CI], 0.93-0.98) (Boice et al., 2003b); this study is limited by imprecise and incomplete data on exposure and information on risk factors.

Ritz (1999) found a weak dose-response relationship with a 15-year lag per 100 mSv of external dose in workers in a uranium-processing plant. Cragle et al. (1988) reported a nonsignificant increase in lung cancer mortality (8 deaths) for salaried and hourly nuclear-fuels production-plant workers (SMR 152) but lower SMRs (also nonsignificant) for only hourly or only salaried workers. The study lacks exposure data. Pinkerton et al. (2004) reported a statistically nonsignificant increase in lung cancer mortality among uranium millers (SMR, 113; 95% CI, 89-141, compared to US referent rates) that was not found in earlier studies of this cohort. When compared to regional referent rates, the increase reached statistical significance (SMR, 151; 95% CI, 119-189). This study is limited by lack of assessment of individual exposure to uranium and other substances in the milling environment.

In summary, there is no consistent evidence of an effect of exposure to natural or depleted uranium on lung-cancer incidence in the studies reviewed. The finding is unchanged when one considers evidence from the studies with the strongest designs, for example, with measurement of cumulative exposure at the individual level, internal controls, a large study population, long followup, and controlling for confounders. The pattern among studies is varied: some studies show increases in risk of lung cancer, and others show decreases. A major shortcoming of the studies is the lack of individual data on smoking, a primary risk factor for lung cancer.

The committee concludes that there is inadequate/insufficient evidence to determine whether an association between exposure to uranium and lung cancer exists.

This conclusion on lung cancer differs from the one in *Volume 1*. The previous committee concluded that there is limited/suggestive evidence of no association between exposure to uranium and lung cancer at cumulative internal doses lower than 200 mSv and that there is inadequate/insufficient evidence to determine whether an association between exposure to uranium and lung cancer exists at higher cumulative exposure (> 200 mSv). The present committee did not place quantitative limits on the dose for the following reasons:

- There is substantial uncertainty in the measurement of uranium exposure in the studies reviewed.
- The types of quantitative measure vary widely from study to study, from individual biomonitoring data to external or internal exposure measurements

(often lacking data on many study subjects) to group estimates based on job title to a general category of years of employment. Furthermore, different dose-reconstruction methods were used to estimate dosage, and different cut-points were often used to categorize the dose in the data analysis, so it was difficult to draw a conclusion.

- Some studies of lung cancer that reported dose had small samples and often did not adjust for risk factors, such as smoking.

Because inhaled uranium dust remains in lung tissues and hilar lymph-node tissues for several years, they are potential targets for uranium radiation. Furthermore, lung cancer is a common malignancy and the leading cause of cancer death; even a modest effect could result in a meaningful increase in the number of cases of lung cancer (that is, an increase in an exposed group compared to an unexposed group might be detectable given the frequency of lung cancer occurrence). Therefore, the committee assigns high priority to continuing to monitor a possible association between exposure to depleted uranium and lung cancer.

Leukemias

The results of only one of the 23 studies reviewed by the committee achieved statistical significance: a residential study by Boice et al. (2003b) (see Table 8-3). The authors reported a reduction in mortality from leukemia (RR [computed by comparing SMRs from the study counties with control counties], 0.91; 95% CI, 0.86-0.97). However, that study is limited by a lack of exposure data and information on other risk factors. The remaining 22 studies showed both increases and decreases in risk associated with exposure to uranium, all of which were nonsignificant. There was no consistent evidence of effect, and the pattern among studies was highly varied. The same pattern was observed after restriction of consideration to the "larger studies" (those with a sample population of about 10,000 or more or with more than 10 cases).

The committee concludes that there is inadequate/insufficient evidence to determine whether an association between exposure to uranium and leukemias exists.

Leukemia is a relatively uncommon malignancy, so large study populations are generally needed to demonstrate any significant moderate effects. The studies reviewed by the committee generally did not have adequate sample size. Earlier studies were complicated by the broad grouping of and changes in classification for leukemia. On the basis of the evidence to date, the committee would assign a low priority to additional study of an association between exposure to depleted uranium and leukemias.

Lymphomas

This section includes discussion of two types of lymphoma: Hodgkin lymphoma (also known as Hodgkin's disease) and non-Hodgkin lymphoma (NHL). The risk of lymphatic malignancy is of particular interest because uranium is known to accumulate in lymph-nodel tissues. Study results are summarized in Tables 8-4 and 8-5.

Hodgkin Lymphoma

The studies considered (see Table 8-4), split virtually evenly between showing an increase in risk of Hodgkin lymphoma associated with exposure to natural or depleted uranium and showing no change or a decrease in risk of Hodgkin lymphoma associated with uranium exposure. The same pattern was observed after restriction of consideration to the "larger studies" (those with a sample population of about 10,000 or more or with more than 10 cases). Only the study by Nuccetelli et al. (2005) achieved a statistically significant finding, showing a significant increase in the risk of Hodgkin lymphoma. Most of the smaller studies show nonsignificantly decreased risk of incidence or death.

Non-Hodgkin Lymphoma and Other Lymphatic Cancers

Table 8-5 presents the results of 24 published studies of a possible relationship between exposure to natural or depleted uranium and NHL. Most of them showed that exposed subjects experienced a risk of NHL equal to or lower than that in unexposed subjects. The same is true if one considers only the larger studies. One study indicated a significant increase in risk: the study by Archer et al. (1973), which had a sample size of only 662, including four cases of lymphatic cancer.

The committee concludes that there is inadequate/insufficient evidence to determine whether an association between exposure to uranium and lymphomas exists. This conclusion applies to both Hodgkin lymphoma and non-Hodgkin lymphoma.

On the basis of the available evidence, the committee concludes that there is a lack of strong and consistent evidence of an association between uranium exposure and lymphatic cancers. The finding is unchanged when one considers evidence from the studies with larger samples and stronger designs: there is no consistent evidence of effect. The pattern among studies is highly varied, as one would expect if there truly were no effect in the population. Although the available evidence does not justify further consideration of a possible association between depleted uranium and lymphatic cancers, the committee concludes that further study of this type of cancer may be warranted on biologic grounds, given that uranium is known to accumulate in the lymph nodes.

Bone Cancer

Twelve studies of uranium-processing workers, one study of a deployed population, and two residential studies assessed bone-cancer outcomes. In most of the studies, the risk of bone cancer was the same or decreased after exposure to natural or depleted uranium (see Table 8-6). Only one study had a significant finding: a statistically significant increase in bone-cancer incidence—four cases—in a Danish military population deployed to the Balkans (SIR, 600; 95% CI, 160-1,530) (Storm et al., 2006). However, because three of the four cases occurred within the first year after deployment, it is unlikely that deployment-related exposure was a factor, given the latency of cancer. After lagging 1 year after deployment, bone-cancer incidence dropped to one case, with a nonsignificant SIR of 170 (95% CI, 0-1,010).

The committee concludes that there is inadequate/insufficient evidence to determine whether an association between exposure to uranium and bone cancer exists.

Overall, the available studies do not provide clear and consistent evidence of an association between natural or depleted uranium and bone cancer. The estimated effects vary greatly from study to study, showing decreased risk, the same risk, or higher risk after exposure. Given that bone cancer is a relatively uncommon malignancy, relatively large study populations are generally needed to demonstrate any significant moderate effects. The studies reviewed by the committee generally did not have adequate sample size. On the basis of the available evidence, the committee would assign a low priority to additional study of an association between exposure to depleted uranium and bone cancer.

Renal Cancer

The committee considered 20 studies of an association between natural or depleted uranium and renal cancer. None of the published results demonstrated a significant increase in risk after uranium exposure (see Table 8-7). The reported SMRs, SIRs, and RRs varied above and below unity except for one residential study (Boice et al., 2003c), which indicated a statistically significant decrease in renal-cancer mortality associated with uranium exposure (RR, 0.58; $p < 0.05$). That study did not include exposure assessment or information on other risk factors. In a more detailed analysis, Dupree-Ellis and colleagues (2000) examined a possible dose-response relationship and found an increasing trend, driven primarily by four renal-cancer deaths in the highest-dose group (excess risk, 10.5/mSV; 90% CI, 0.6-57.4). That result was not statistically significant.

The committee concludes that there is inadequate/insufficient evidence to determine whether an association between exposure to uranium and renal cancer exists.

None of the 20 studies considered by the committee demonstrated a significant increase in risk of renal cancer after exposure to uranium. When attention was restricted to the studies with the largest samples, there was no positive evidence of an effect at the low exposures observed in the studies. On the basis of the available evidence, the committee would assign a low priority to further study of an association between exposure to depleted uranium and renal cancer.

Bladder Cancer

The committee evaluated 20 published studies of a potential association between exposure to natural or depleted uranium and bladder cancer: 14 uranium-processing studies, two studies of military populations, and four residential studies (see Table 8-8). Most of the studies reported the same or reduced bladder-cancer mortality or incidence in exposed subjects. Only one finding achieved statistical significance: a UK processing study found a significant reduction in bladder-cancer incidence (SIR, 76; $p < 0.05$) but roughly equal mortality (SMR, 92; nonsignificant) (McGeoghegan and Binks, 2000b). That study is limited by a lack of data on internal radiation exposure and other risk factors. Two studies of veterans deployed to the Balkans reported increased but nonsignificant SIRs for bladder cancer, but both studies were based on very small numbers of observed cases (Gustavsson et al., 2004; Storm et al., 2006).

The committee concludes that there is inadequate/insufficient evidence to determine whether an association between exposure to uranium and bladder cancer exists.

Overall, the committee finds little evidence that exposure to natural or depleted uranium increases the risk of bladder cancer. Most of the studies, whether small or large, show the same or reduced risk of bladder cancer in people exposed to uranium. Although the two studies of deployed populations showed nonsignificant increases in risk, the estimates were based on small numbers of cases—two and seven. A small number of cases renders findings less robust in that changes in exposure or outcome status in only one or two people could have altered the findings substantially, so confidence in the findings is reduced. The committee would assign a low priority to further study of an association between exposure to depleted uranium and bladder cancer.

Brain and Other Central Nervous System Cancers

Findings of 20 published studies of an association between uranium exposure and brain and other central nervous system cancers are described in Table 8-9. Almost all failed to demonstrate statistically significant associations between uranium exposure and brain and other central nervous system cancers, but they are roughly evenly split between those showing increases in and those showing the same or decreases in mortality or incidence. That overall pattern is unchanged if one restricts attention to the larger or better designed studies. Only two studies had significant results: significant decreases in risk after uranium exposure. The study by Cragle et al. (1988) reported a statistically significant decrease in mortality after exposure in hourly workers at a nuclear-fuels production facility (SMR, 23; $p < 0.05$). However, the SMRs for salaried workers and for combined hourly and salaried workers were not statistically significant. In addition to a possible healthy-worker effect, the study may be limited by a lack of detailed exposure assessment and the use of "hourly" vs "salaried" as a proxy for socioeconomic status. Beral et al. (1988) also reported a significant deficit in mortality from brain and other nervous system cancers in processing workers (SMR, 32; $p < 0.05$).

The committee concludes that there is inadequate/insufficient evidence to determine whether an association between exposure to uranium and cancers of the central nervous system, including brain cancer, exists.

The published studies show inconsistent results that do not lead to a conclusion of an association between natural or depleted uranium and cancers of the central nervous system. Studies of some other cancers (for example, bladder cancer) showed an equal or reduced risk after exposure, but the distribution of studies of brain and other central nervous system cancers is more balanced: results are roughly equally divided between studies that show increased risk and studies that show the same or decreased risk. Because of that pattern, the committee believes that further study of an association between depleted uranium and central nervous system cancers may be warranted but should not be assigned a high priority.

Stomach Cancer

The committee considered 21 published studies of a possible association between natural or depleted uranium and stomach cancer, including 16 processing studies, one study of military populations, and four residential studies (see Table 8-10). All but three had statistically nonsignificant results, and most demonstrated the same or decreased mortality or incidence. The pattern is unchanged if one restricts consideration to the larger or better designed studies. The three studies that had statistically significant results all showed a decrease in mortality or incidence (Beral et al., 1988; Dupree-Ellis et al., 2000; McGeoghegan and

Binks, 2000b). McGeoghegan and colleagues found a significantly decreased risk of stomach cancer (SIR, 76; $p < 0.05$) but an approximately equal risk of stomach-cancer death (SMR, 92; nonsignificant) in workers at the Springfields uranium-production facility (McGeoghegan and Binks, 2000b); however, the study is limited by inadequate data on exposure, particularly internal exposure.

> **The committee concludes that there is inadequate/insufficient evidence to determine whether an association between exposure to uranium and stomach cancer exists.**

Overall, the committee finds little evidence to suggest that exposure to natural or depleted uranium increases the risk of stomach cancer. Most of the studies showed similar or reduced risk of stomach-cancer death and incidence in people exposed to uranium. Although four uranium-processing studies showed nonsignificant increase in SMRs, the findings were based on 15 or fewer cases. Similarly, the study of Danish deployed populations that showed a nonsignificant increase in risk was based on two cases. Therefore, confidence in the findings is low. In the view of the committee, further study of an association between depleted uranium and stomach cancer would have a low priority.

Male Genital Cancers

Prostatic cancer is the most frequently diagnosed cancer in men in the United States, and any increase in risk could result in a large increase in the number of cases or deaths. Testicular cancer, the most common cancer among young men, is of special interest to Gulf War veterans, and some studies of veterans suggested a higher but nonsignificantly increased risk (IOM, 2006).

Prostatic Cancer

The committee evaluated 19 published studies of a potential association between exposure to natural or depleted uranium and prostatic cancer, including 14 processing studies, two studies of deployed populations, and three residential studies (see Table 8-11). Only one reported a statistically significant finding: McGeoghegan and Binks (2000b) found a significant reduction in prostatic-cancer incidence (SIR, 77; $p < 0.05$) but not mortality (SMR, 89; nonsignificant) in workers at the Springfields uranium-processing plant. The study is limited by the lack of data on internal radiation exposure. Three other studies of processing workers reported increased prostatic-cancer mortality, but none of the SMRs was statistically different from the null value indicating no effect (Beral et al., 1988; Loomis and Wolf, 1996; Ritz, 1999).

The larger studies (those with samples of about 10,000 or more or with more than 10 affected cases) had more findings of decreased risk than of increased risk in those exposed to uranium. No study showed a statistically significant increase in risk. The only statistically significant finding was a decrease in cancer incidence (SIR, 77; $p < 0.05$). Overall, there is little evidence of an association between uranium exposure and prostatic cancer.

The committee concludes that there is inadequate/insufficient evidence to determine whether an association between exposure to uranium and prostatic cancer exists.

Of the 19 studies considered, none demonstrated a significantly increased risk of prostatic cancer after exposure to uranium, and one showed a significant decrease in cancer incidence but not mortality. If only the studies with the largest samples are considered, the committee finds that there is no affirmative evidence of effect. On the basis of the available evidence, the committee would assign a low priority to further study of an association between exposure to depleted uranium and prostatic cancer.

Testicular Cancer

Table 8-12 summarizes the findings of 15 published studies considered by the committee for a possible relationship between exposure to natural or depleted uranium and testicular cancer, including 11 studies of uranium-processing workers, three studies of military populations, and one study of residents living near a nuclear facility in Pennsylvania. None of the results achieved statistical significance. All studies of processing workers showed reduced testicular-cancer mortality in people exposed to uranium but did not reach the 5% level of statistical significance. All three studies of deployed veterans found increased incidence rate ratios or SIRs, but they also did not reach statistical significance (Macfarlane et al., 2003; Gustavsson et al., 2004; Storm et al., 2006).

The committee concludes that there is inadequate/insufficient evidence to determine whether an association between exposure to uranium and testicular cancer exists.

The committee finds no consistent evidence that uranium exposure increases the risk of testicular cancer. All occupational cohorts had lower mortality. Testicular cancer, although very rare in the general population, is common in young adults and therefore prevalent in deployed veterans. The nonsignificant excess in incidence observed in the studies of military populations could be due in part to routine medical surveillance of the deployed veterans. Despite the inconsistent evidence, testicular cancer is of special interest to Gulf War veterans. The committee believes that further study of an association between depleted uranium and testicular cancer may be warranted but should not be assigned a high priority.

Other Cancers

A study of health outcomes in 53,462 Gulf War veterans reported only all-cancer incidence, not site-specific incidence (Macfarlane et al., 2005). It did not find a statistically significant increase in cancer incidence (mortality rate ratio, 1.01; 95% CI, 0.79-1.30). However, the 13-year followup period may be too short for most cancers to have developed.

Early studies by Archer et al. (1973), Wagoner et al. (1964), and Waxweiler et al. (1983) combined hematopoietic and lymphopoietic cancers, but only one (that by Archer et al.) found a significant increase (SMR, 392; $p < 0.05$). Beral et al. (1988) also found a significantly lower RR of all lymphopoietic and hematopoietic cancers (RR, 0.46; 95% CI, 0.23-0.94) in workers with radiation-exposure records than in those without exposure records.

NONCANCER OUTCOMES

The following subsections present the strength of the evidence of associations between exposure to natural or depleted uranium and specific nonmalignant health outcomes. They draw on the information from the many studies that were described in Chapter 7 and *Volume 1*. The committee has highlighted the relevant findings on nonmalignant outcomes from the literature, with a focus on outcomes related to the organs and organ systems likely to be affected by natural or depleted uranium, such as the kidneys and the respiratory, central nervous, and reproductive systems. The findings show both positive and negative associations between uranium and nonmalignant health outcomes.

Nonmalignant Renal Disease

Mortality

Fourteen studies assessed the association between occupational exposure and renal-disease mortality. Four reported an excess in mortality that was not statistically significant (see Table 8-13). Two of those followed the mortality experience of uranium millers in the Colorado Plateau region. In 1983, Waxweiler and colleagues reported an excess in deaths from chronic nephritis (SMR, 167; 95% CI, 60-353). However, all deaths in the group occurred in short-term workers, and this lessened the likelihood that the deaths were related to uranium exposure (IOM, 2000). In a followup study of the Colorado group, Pinkerton and colleagues also observed an increase in mortality due to chronic renal disease (SMR, 135; 95% CI, 58-267) (Pinkerton et al., 2004) that was not statistically significant. Similarly, Dupree-Ellis and colleagues (2000) found an excess in mortality from chronic nephritis (SMR, 188; 95% CI, 75-381) in workers at the Mallinckrodt Chemical works plant that was not statistically significant. The authors noted that prior exposure to silica in previous jobs and misclassification of renal diseases may have limited the interpretability of their results. McGeoghegan and Binks (2001) found a nonsignificant increase in deaths due genitourinary diseases in radiation workers compared with the English and Welsh populations (5 observed vs 4.63 expected; SMR, 108) in a study of processors at the British Nuclear Fuels Chapelcross site.

Cragle and colleagues (1988) reported statistically significantly fewer deaths due to genitourinary diseases in hourly employees (SMR, 39; 95% CI, 10-96) in a study of workers at the Savannah River plant. McGeoghegan and Binks (2000b) also reported significantly fewer deaths than expected in radiation workers (SMR, 57; $p < 0.01$). Frome and colleagues (1997) reported fewer deaths than expected from diseases of the genitourinary system (SMR, 83) in white men in a study of processing workers at the four Federal nuclear plants in Oak Ridge, Tennessee. An earlier study of Oak Ridge workers at the Y-12 and K-25 uranium-enrichment facilities revealed no difference between the numbers of observed and expected deaths from chronic nephritis (SMR, 99; 95% CI, 71-126[1]) (Frome et al., 1990), as reported in *Volume 1*. The observed findings were probably influenced by a healthy-worker effect.

In many of the 14 studies, the computed death rates included all genitourinary conditions instead of focusing on renal diseases. Despite reported increases in observed deaths, the SMRs may not have reflected a true response to uranium exposure. In several of the plants, uranium exposure coexisted with other relevant heavy-metal or chemical exposure. Generally, most researchers were unable to isolate the effects of uranium exposure alone.

Morbidity

Gulf War Veterans Depleted-Uranium Surveillance Study

McDiarmid and colleagues conducted a medical investigation of Gulf War veterans who inhaled or ingested airborne depleted-uranium particles or experienced depleted-uranium wound contamination as a result of friendly-fire incidents and found renal-function measurements that were generally within normal clinical limits (see Table 8-14) (McDiarmid et al., 2000, 2001, 2004, 2006, 2007). Urinary uranium excretion was used in the exposure assessment, and subjects were separated into high- and low-exposure groups on the basis of a cutpoint of 0.10 µg/g of creatinine. In the first of the Baltimore Veterans Affairs Medical Center (BVAMC) studies, veterans with retained depleted-uranium shrapnel fragments had higher urinary uranium concentrations than those without 7 years after first exposure. Urinary uranium ranged from 0.01 to 30.74 µg/g of

[1]The confidence interval was calculated by the Committee on Health Effects Associated with Exposure During the Gulf War; it was not stated in the original study (IOM, 2000).

creatinine in veterans with retained fragments and 0.01 to 0.05 μg/g creatinine in veterans without fragments. Despite that finding, renal-function measures (serum creatinine, beta-microglobulin, retinol-binding protein, serum uric acid, urinary creatinine, and urinary protein) were quite similar between the high- and low-exposure groups (McDiarmid et al., 2000). In the 1999 evaluation, urinary uranium ranged from 0.018 to 39.1 μg/g of creatinine in the depleted-uranium–exposed veterans with retained fragments and 0.002 to 0.231 μg/g of creatinine in depleted-uranium–exposed veterans without fragments. Clinical tests revealed renal measures within normal limits with slight differences between high- and low-uranium groups. The authors did not detect any clinically important changes in renal function due to depleted-uranium exposure; urinary creatinine concentration was slightly lower in the high-uranium group, but the difference was only marginally significant (McDiarmid et al., 2001). An increase in urinary uranium (24-hour urinary uranium concentrations higher than 0.05 μg/g of creatinine) was seen in four of the 30 newly enrolled veterans (McDiarmid et al., 2002). The 2001 surveillance reported urinary uranium ranging from 0.001 to 78.125 μg/g of creatinine. The presence of retained depleted-uranium shrapnel appeared to be associated with higher urinary uranium concentration. In addition, most urinary-uranium results were consistent over time. Mean values of all renal-function markers were within normal clinical limits with few statistically significant differences between high- and low-uranium groups 10 years after first exposure. Serum creatinine was higher in the low-uranium group (0.85 vs 0.95 mg/dL; $p = 0.03$), and urinary retinol-binding protein (65.58 vs 46.13 μg/g of creatinine; $p = 0.06$) and total urinary protein (78.69 vs 54.63 mg/g of creatinine; $p = 0.01$) were higher in the high-uranium group (McDiarmid et al., 2004). Those differences were not observed in the previous evaluations of this group. In 2003, all but one of the renal measures were within normal clinical limits. The difference in serum phosphate concentration was the only measurable difference between the high- and low-exposure groups (4.11 vs 3.75 mg/dL; $p = 0.03$) (McDiarmid et al., 2006), but its clinical importance is unclear.

In the most recent evaluation, urinary uranium ranged from 0.002 to 44.1 μg/g of creatinine in total 24-hour urine, and participants with known embedded depleted-uranium shrapnel fragments and specific uranium indicators of depleted uranium had concentrations at or above the cutpoint of 0.10 μg/g of creatinine. The results showed a high correlation between current and cumulative uranium-exposure measures. Of the 34 veterans with depleted-uranium shrapnel, 10 had current urinary uranium concentrations that exceeded the cutpoint of 0.10 μg/g of creatinine. The same number had cumulative urinary uranium concentrations over the cutpoint of 10 μg/g of creatinine. Differences in mean serum uric acid were borderline ($p = 0.03$) when groups with high and low cumulative uranium exposure were compared. Despite that finding, the values were within the normal clinical range, and the differences were small: 5.22 mg/dL in the high group and 6.19 mg/dL in the low group. Other renal characteristics had no significant differences whether current or cumulative uranium measures were used (McDiarmid et al., 2007).

Drinking Water and Residential Exposure

Kurttio and colleagues investigated renal measures related to uranium exposure through drinking water in 325 Finnish people who obtained their water from drilled wells (see Table 8-14). The 2002 report on the cohort noted a statistically significant association between uranium exposure and calcium excretion ($p = 0.03$) in well-water users (Kurttio et al., 2002). The authors documented an association between urinary uranium and fractional excretion of calcium for all exposure metrics. They also observed a statistically significant association between urinary uranium and fractional phosphate ($p = 0.03$). There was no association between uranium exposure and measures of glomerular function (Kurttio et al., 2002).

In a later study of the cohort, Kurttio and colleagues (2006a) found that urinary uranium concentrations were an average of 44% greater than during prior sampling. The study further examined renal toxicity due to uranium exposure through drinking water in 193 of the 325 people included in the 2002 study. In general, markers of renal function were within normal limits. Biomarkers of cytotoxicity, renal proximal tubular function, glomerular function, and other exposure indicators were not significantly associated with urinary uranium concentration. However, there were statistically significant associations between cumulative uranium intake and glucose excretion ($p = 0.02$) and between uranium exposure and increased blood pressure (diastolic, $p = 0.01$; systolic, $p = 0.07$).

In the only study that examined an association between residential exposure and renal effects, researchers observed a statistically significant excess in renal disease (standardized prevalence ratio [SPR], 215; 99% CI, 186-248) and bladder disease (SPR, 132; 99% CI, 111-156) in people who lived near the Fernald Feed Materials Production Center (FFMPC) in Ohio. The outcomes included increases in a few subcategories, such as kidney stones (SPR, 398; 99% CI, 336-468) and chronic nephritis (SPR, 203; 99% CI, 76-435). However, the health outcomes were self-reported, and some were not verified, so the potential for outcome misclassification was increased. Residents who obtained their drinking water from a well or cistern had higher urinary microalbumin concentrations (Pinney et al., 2003).

Occupational Uranium Exposure

Boiano and colleagues conducted a medical investigation of workers at the FFMPC (see Table 8-14). They observed urinary uranium concentrations up to 13 μg/L, and 109 of the participants had concentrations under the detection limit of 5 μg/L. However, no associations were observed between measures of uranium exposure and glomerular filtration or tubular markers (Boiano et al., 1989).

A study of processors in Egypt (Shawky et al., 2002) found a mean urinary uranium concentration of 17.8 μg/L. It also reported that urinary uranium was increased in the 13 participants who provided spot urine specimens, ranging from 8 to 29 μg/L. There was a correlation between urinary uranium and serum creatinine in the 13 specimens, and mean uranium excretion was more than 20 times the occupational-exposure decision level of 0.8 μg/L. However, there were no individual exposure data other than data on the 13. That, in addition to the small sample and the absence of more specific markers for evaluating tubular dysfunction, limits the value of the reported results.

Conclusion

Although high exposure to uranium, a heavy metal, is known to be toxic to the kidneys (see Chapter 3 for a discussion of the toxicity of uranium in animal models), the literature evaluated does not provide substantial evidence of an association between exposure to natural or depleted uranium and important clinical renal effects in humans. Several studies found slight changes in renal markers but no abnormalities in renal function. Gulf War veterans exposed to depleted uranium in embedded shrapnel had minor changes in renal measures and increased urinary uranium concentrations over the course of a 14-year followup, but overall mean values remained within normal clinical ranges. The studies of well-water users in Finland thoroughly characterized the nature of exposure but examined renal effects in a small group and included a relatively short followup. Studies of workers in processing plants at the Fernald Feed Materials Production Center detected no association between uranium exposure and glomerular or tubular markers.

The committee concludes that there is inadequate/insufficient evidence to determine whether an association between exposure to uranium and nonmalignant renal disease exists.

This conclusion on renal disease differs from the one in *Volume 1*. The previous committee concluded that there is limited/suggestive evidence of *no* association between exposure to uranium and clinically significant renal dysfunction. On the basis of the available evidence, the present committee could not rule out renal effects after exposure of any magnitude (see Chapter 4 for the definition of the category of limited/suggestive evidence of *no* association). The committee also could not place quantitative limits on the dose, for reasons similar to those detailed above in connection with lung cancer.

The published research evidence is inadequate to support a conclusion about depleted uranium as a cause of nonmalignant renal disease. The well-observed renal effects of heavy metals that are excreted in urine make a deleterious effect of depleted-uranium exposure plausible if the exposure is of sufficient magnitude and duration. The kidneys are identified as the most sensitive target of uranium toxicity in the US Army's "Capstone Report" (USACHPPM, 2004) and in the National Research Council report, *Review of Toxicologic and Radiologic Risks to Military Personnel from Exposure to Depleted Uranium During and After Combat* (NRC, 2008). However, available modes of uranium exposure—industrial exposure, groundwater exposure, and depleted-uranium exposure of a small number of veterans—do not indicate renal toxicity in these settings. Additional studies of larger numbers of exposed people with well-characterized exposure and renal outcomes will be needed before any definitive conclusions can be drawn about a nephrotoxic effect of exposure to depleted uranium in a war theater. On the basis of the available evidence, the committee would assign a high priority to further study of an association between exposure to depleted uranium and nonmalignant renal disease.

Nonmalignant Respiratory Disease

The committee evaluated 14 mortality and two morbidity studies of exposure to uranium and nonmalignant respiratory disease (see Tables 8-15 and 8-16). In a 2004 study of a cohort of uranium millers in the Colorado Plateau, Pinkerton and colleagues (2004) observed a significant increase in mortality from nonmalignant respiratory disease compared with the US referent population (SMR, 143; 95% CI, 116-173) due to an excess in mortality from emphysema (SMR, 196; 95% CI, 121-299) and pneumoconioses and other respiratory diseases (SMR, 168; 95% CI, 126-221). Those findings were consistent with those of a previous study of the cohort (Waxweiler et al., 1983). However, mortality from emphysema was higher in workers employed before 1955, when exposures to silica and vanadium, in addition to exposure to uranium, were thought to be at their highest (before 1995: 17 observed; SMR, 222; 95% CI, 129-356; 1955 or later: 4 observed; SMR, 130; 95% CI, 36-333) (Pinkerton et al., 2004). Frome and colleagues (1990) also reported a significant excess in deaths from nonmalignant respiratory diseases. However, several studies found decreases in lung-disease mortality. As reported in *Volume 1*, Ritz (1999) found a significant decrease based on 53 deaths. As with mortality from nonmalignant renal diseases, the respiratory-disease outcomes were grouped, so the ability to observe effects of individual diseases was reduced. In addition, the issue of exposure to multiple respiratory toxicants is important with respect to respiratory disease: many workers were often exposed to other agents (such as silica) known to have effects on the lungs.

In a study of lung disease in workers at the FFMPC, investigators found some associations between indicators of uranium exposure and respiratory effects. The ratio of 1-second forced expiratory volume (FEV_1) to forced vital capacity was associated with the job-history–derived uranium-exposure index after adjustment for smoking. However, the FEV_1 alone was not associated with the exposure index. Shortness of breath was significantly associated with self-reported uranium exposure (Boiano et al., 1989). People who lived close to the plant had significantly fewer cases of asthma (SPR, 85; 99% CI, 73-98), chronic bronchitis (SPR, 19; 99% CI, 14-24), and emphysema (SPR, 61; 99% CI, 41-68) compared with National Health Interview Survey rates (Pinney et al., 2003).

The committee concludes that there is inadequate/insufficient evidence to determine whether an association between exposure to uranium and nonmalignant respiratory disease exists.

Results of several of the studies support an effect of employment in uranium-processing facilities on nonmalignant respiratory disease, but their applicability to military depleted-uranium exposure is limited by the extent of concomitant coexposure of such workers to other respiratory toxicants (such as silica, asbestos, and vanadium). Results of inhalation studies of various forms of uranium in several animal species are inconsistent with respect to nonmalignant respiratory effects (see Chapter 3). On the basis of the available evidence, the committee would assign a high priority to further study of an association between exposure to depleted uranium and nonmalignant respiratory disease.

Neurologic Effects

The studies of uranium-processing workers showed no excess in neurologic-disease mortality (Polednak and Frome, 1981; Cragle and et al., 1988; Frome et al., 1990, 1997; Dupree-Ellis et al., 2000; McGeoghegan and Binks, 2000a,b, 2001; Boice et al., 2006) (see Table 8-17). As part of the Depleted Uranium Follow-up Program at the BVAMC, McDiarmid and colleagues used various traditional and automated test batteries (see Chapter 7) to assess neurocognitive performance in veterans. Results of the evaluation of Gulf War veterans suggested a statistically significant relationship between increased urinary uranium concentrations and poor performance on automated neuropsychologic tests regardless of the models used (24-hour-urine uranium in depleted-uranium–exposed veterans, $p = 0.01$; spot-urine uranium in all veterans, $p = 0.01$); traditional test measures showed no statistical differences between exposed and unexposed veterans (McDiarmid et al., 2000). However, the relationship between urinary uranium concentration and performance on automated measures observed in the 1994 and 1997 evaluations appeared to weaken and had only a marginal level of significance ($p = 0.098$) in high and low urinary-uranium groups in the 1999 surveillance after adjustment for intelligence (WRAT-3) and depression (Beck Depression Inventory) (McDiarmid et al., 2001). Later surveillance (2001, 2003, and 2005) found no statistically significant differences between exposure groups in neurocognitive indexes (McDiarmid et al., 2004, 2006, 2007). A modest association was seen between urinary uranium and the accuracy impairment (A-IIac) index in 2001 and 2003 surveillance, but the authors noted that the result was based on test performance of two veterans whose uranium concentrations were exceedingly high.

The committee concludes that there is inadequate/insufficient evidence to determine whether an association between exposure to uranium and nonmalignant diseases of the nervous system exists.

Overall, published studies of neurologic outcomes are either negative studies that do not find any evidence of health effects of exposure to depleted uranium or relatively small studies, such as the Depleted Uranium Follow-up Program at the BVAMC, that find inconstant associations. As described in Chapter 3, the results of studies in animal models indicate that depleted uranium is a toxicant capable of crossing the blood-brain barrier. Data on effects are inconsistent; some animal studies report behavioral changes, and others do not. Although at high concentrations different forms of uranium might be associated with some subtle neurologic dysfunction in animals, the relevance of these observations to humans remains unknown. On the basis of the available evidence, the committee would assign a high priority to further study of an association between exposure to depleted uranium and neurologic effects.

Reproductive and Developmental Effects

A few studies examined the effects of natural or depleted uranium on human reproduction and development (see Table 8-18). McDiarmid and colleagues evaluated endocrinologic function in Gulf War veterans by measuring blood concentrations of follicle-stimulating hormone (FSH), luteinizing hormone (LH), prolactin, testosterone, thyroid-stimulating hormone, and free thyroxine. Study authors also assessed semen for a number of characteristics, including volume, concentration, structure, and motility. A statistically significant difference was observed in mean prolactin concentrations, which were 1.66 and 12.47 μg/g of creatinine ($p = 0.04$) in low- and high-prolactin groups (McDiarmid et al., 2000).

In the 1999 surveillance, there were no statistically significant differences in mean FSH, LH, prolactin, and testosterone concentrations or thyroid measures between low and high groups. Of the 44 sperm samples included in the analysis, three were designated subnormal—possessing below normal values of at least three of the five characteristics as defined by World Health Organization standards. The high-urinary-uranium groups had more abnormal total sperm counts (583.5 ± 106.1 vs 286.6 ± 44.8), total progressive sperm counts (220.9 ± 44.0 vs 108.2 ± 19.2), and total rapid progressive sperm counts (155.5 ± 31.1 vs 81.3 ± 15.4) that were statistically significant ($p = 0.02$, 0.03, and 0.04, respectively), results not previously seen in this group (McDiarmid et al., 2001). In 2001, overall neuroendocrine function was normal, but mean free thyroxine was higher in the low-uranium group (1.66 vs 1.08 ng/dL), a result not observed in the 1997 and 1999 evaluations. There was no statistically significant difference in semen measures between the high- and low-urinary-uranium groups (McDiarmid et al., 2004); this finding was consistent in successive evaluations. Increased mean values of semen characteristics were seen in the high-urinary-uranium group in the 2003 evaluation, but all were within the normal clinical ranges (McDiarmid et al., 2006). There were no statistically significant differences between high- and low-urinary-uranium groups in neuroendocrine measures, which were generally within normal clinical limits in veterans examined in the following surveillance. Mean values of semen characteristics also showed no statistically significant differences; however, the percentages of progressive sperm and rapid progressive sperm were lower in the high-uranium group on the basis of the current urinary-uranium metric (McDiarmid et al., 2007).

In a study of the prevalence of major malformations in two 1-year cohorts of neonates born in 1995 (immediately after the war in Bosnia) and in 2000 (5 years after military activities), 40 of 1,853 (2.16%) in 1995 had major malformations (95% CI, 1.49-2.82%), and 33 of 1,463 (2.26%) in 2000 had major malformations (95% CI, 1.50-3.01%). In addition, anomalies of the cardiovascular system (0.615% vs 0.162%) and central nervous system (0.273% vs 0%) were more elevated in the 2000 cohort than in the 1995 cohort (Sumanovic-Glamuzina et al., 2003).

The committee concludes that there is inadequate/insufficient evidence to determine whether an association between exposure to depleted uranium and reproductive and developmental effects exists.

Relatively large study populations are generally necessary to demonstrate significant but subtle reproductive or developmental effects. The studies reviewed generally had too few subjects or relied on insufficiently precise exposure assessment to support definitive conclusions. Although some toxicology studies have reported that exposure of animals to uranium compounds during development can lead to a variety of adverse effects, others did not find that uranium exposure affected reproduction and development (see Chapter 3).

On the basis of the available evidence, the committee would assign a high priority to further study of an association between exposure to depleted uranium and reproductive and developmental effects.

Other Health Outcomes

The following discussion of additional health outcomes focuses on reported cardiovascular, hematologic, genotoxic, bone, and immunologic effects of exposure to natural or depleted uranium. The outcomes have not been studied in detail in humans, so the evidence from which to draw conclusions is sparse. The results presented here come primarily from a case series of Gulf War veterans who participated in the Depleted Uranium Follow-up Program at the BVAMC and from studies of uranium-processing workers and well-water users in Finland.

Cardiovascular Effects

Mortality from diseases of the circulatory system was significantly lower in most studies of uranium-processing workers, probably because of the healthy-worker effect. Pinkerton and colleagues reported statistically significantly fewer deaths from heart disease than expected (SMR, 84; 95% CI, 75-94) in a cohort of uranium-mill workers in the Colorado Plateau region (Pinkerton et al., 2004). Similarly, workers employed in the FFMPC had lower cardiovascular mortality than the US white male population (SMR, 78; 95% CI, 71-86) (Ritz, 1999). Mortality from circulatory diseases in workers at the Mallinckrodt processing plant (SMR, 89; 95% CI, 81-97) (Dupree-Ellis et al., 2000) and the Rocketdyne/Atomics International (SMR, 68; 95% CI, 58-78) (Ritz et al., 2000) was significantly lower than that in white males in the United States. Results of experimental studies that used exceedingly high doses of uranium in several animal models suggest that the cardiovascular system is not a sensitive target for this metal.

Genotoxic Effects[2]

McDiarmid and colleagues (2001) found a statistically significant increase in mean sister-chromatid exchanges (SCEs) (6.35 ± 0.267 vs 5.52 ± 0.182; $p = 0.03$) in cultured peripheral-blood lymphocytes from members of the high-urinary-uranium group in the 1999 medical surveillance of depleted-uranium–exposed Gulf War veterans. The association remained after adjustment for current smoking status. A statistically significant difference between low- and high-exposure groups was also seen in mean SCEs at high doses of bleomycin; the high-exposure group had increased SCEs (6.25 ± 0.338 vs 4.88 ± 0.262; $p = 0.01$). No differences were observed in tests for chromosomal aberrations. The findings suggest a possible genotoxic effect; however, as the authors suggest, additional surveillance was needed to establish a clinical association. In the 10-year postwar followup assessment, the authors reported a statistically significant increase in the mean frequency of chromosomal aberrations in the high-urinary-uranium group (McDiarmid et al., 2004). However, the 12- and 14-year assessments revealed no statistical differences in chromosomal aberrations between high- and low-urinary-uranium groups. Hypoxyanthine-guanine phosphoribosyl tranferase mutation frequencies measured at 10, 12, and 14 years were nonsignificantly greater in the high-exposure group than in the low-exposure group.

---

[2]Human genotoxic effects are covered in greater detail in Chapter 4.

Hematologic Effects

In general, hematologic measures in depleted-uranium–exposed Gulf War veterans were within normal clinical limits. Clinical tests revealed slight differences between high- and low-urinary-uranium groups. In a 1999 surveillance of veterans, hematologic measures exhibited statistically significant differences between high- and low-exposure groups. The high-urinary-uranium group had a lower mean lymphocyte count (32% vs 37%; p = 0.04), a higher mean neutrophil percentage (55% vs 49%; p = 0.03), and a lower mean monocyte percentage (7.6% vs 9.1%; p = 0.01) (McDiarmid et al., 2001). Differences in hematocrit (42.59% in the high-uranium group and 44.60% in the low-uranium group) and hemoglobin (14.79 vs 15.40 g/dL) that were not observed in the 1997 and 1999 surveillance were seen in 2001 (McDiarmid et al., 2004). The most recent evaluation found no statistically significant differences between high- and low-urinary-uranium groups in hematologic and blood-chemistry measures; they were within normal clinical limits (McDiarmid et al., 2007). Overall, increased urinary uranium excretion had little effect on hematologic measures.

Immunologic Effects

Only one study examined immunologic effects of depleted uranium. McDiarmid and colleagues found a significantly higher proportion of CD4+ T cells in the high- than in the low-uranium group (65.98% vs 60.83%), and CD8+ T cells were significantly lower in the high- than in the low-uranium group (26.55% vs 31.28%) (McDiarmid et al., 2004).

Skeletal Effects

In studies of the effect of uranium exposure on bone, researchers focused on biochemical markers of bone resorption and formation. In a study of Finnish well-water users, uranium exposure was shown to be associated with increased CTx (a bone-turnover marker) in men (uranium in water, p = 0.05 and 0.01; daily intake, p = 0.16 and 0.02; and cumulative intake, p = 0.16 and 0.03, in the robust and linear-regression analyses, respectively). In addition, uranium concentrations in drinking water appeared to be associated with increased osteocalcin, a biomarker often used for bone formation (p = 0.19; p = 0.04 in linear-regression analysis). Uranium exposure was not related to any biomarkers of bone metabolism in women. Amino-terminal propeptide of type I procollagen was not associated with uranium exposure (Kurttio et al., 2005). In an analysis of tissue collected during an autopsy of a uranium-processing worker, uranium was found to be deposited more in bone than in the liver or kidneys (Kathren et al., 1989).

Conclusion

The committee concludes that there is inadequate/insufficient evidence to determine whether an association between exposure to uranium and cardiovascular, genotoxic, hemotologic, immunologic, and skeletal effects exists.

SUMMARY

This chapter summarized the committee's systematic evaluation of the scientific literature about the human health outcomes of exposure to uranium. Overall, the committee concluded that the available data are inadequate and insufficient to support statements that exposure to uranium is associated with the health outcomes or statements that exposure to uranium is *not* associated with the health outcomes. The inability to reach positive or negative conclusions is due largely to limitations of the available scientific literature. Studies that permit more definitive conclusions might become available in the future.

The committee's review and evaluation of the scientific literature placed particular emphasis on epidemiologic studies. Toxicologic data were considered secondary and were used largely to determine mechanism of action. The committee used direct evidence (that is, from the empirical literature) rather than relying on a theory-driven approach (that is, using mechanistic models) in drawing its conclusions.

Most of the evidence on health outcomes of exposure to uranium comes from studies of workers in uranium-processing mills and other facilities, and the committee relied heavily on those studies in developing its conclusions. It also considered studies of Gulf War veterans who were exposed to depleted uranium and studies of residential exposure to uranium. The committee selected studies that it believed to be the most relevant to identifying health outcomes in depleted-uranium–exposed military personnel. Although numerous epidemiologic studies of various forms of radiation exposure have been conducted, the committee limited its review to studies of exposure to uranium (both natural and depleted uranium).

The use of the epidemiologic literature in developing conclusions presented several limitations. For example, the number of exposed people in many of the studies was relatively small, and this decreased the statistical power to detect small excesses of disease. The period of followup in several studies might have been too short to detect some diseases that are typically characterized by long latency; this limitation is of particular concern in regard to studies of cancer outcomes. Appropriate classification of study subjects according to exposure status also constituted a limitation. Inaccurate or imprecise characterization of the exposure of each person in a study may reduce the likelihood of detecting a health outcome associated with exposure or, conversely, could lead to the appearance of an association when none exists. Assessment of exposure to uranium was inadequate in many of the studies reviewed by the committee.

The likelihood of detecting an association between exposure and a health outcome depends on several factors (see Chapter 4). For the health outcomes discussed in this chapter, the committee concluded that exposure to uranium is not associated with a large or frequent effect. Nevertheless, it is possible that depleted-uranium–exposed veterans will have a small increase in the likelihood of developing a disease. Typically, extremely large study populations are necessary to demonstrate that a specific exposure is not associated with a health outcome. The committee's evaluation of the literature supports the conclusion that a large or frequent effect is unlikely, but it is not possible to state conclusively that a particular health outcome cannot occur.

In summary, the committee assigned the category *inadequate/insufficient evidence to determine whether an association exists* to each health outcome described above for one or more of the following reasons:

- Well-conducted studies showed equivocal results.
- The magnitude or frequency of a health outcome may be so low that it cannot be reliably detected given the sizes of the study populations.
- The available studies had limitations (for example, inadequate exposure assessment or followup that was too short) that prevented the committee from reaching clear conclusions about health outcomes.

For those reasons, a conclusion of inadequate/insufficient evidence to determine whether an association exists is not synonymous with evidence of no association.

TABLE 8-1 Lung Cancer

Study	Cohort/Study Site	Population	No. Observed Cases/Deaths	No. Expected Cases/Deaths	SMR	95% CI
Studies of Uranium Processors						
Wagoner et al., 1964	Uranium mills, Colorado Plateau	611	0 (all respiratory cancers)	1.9	0	NS
Archer et al., 1973	Uranium mills, Colorado Plateau	622	4 (all respiratory cancers)	4.26	94	NS (–3 to 191, from *Volume 1*)
Waxweiler et al., 1983	Uranium mills, Colorado Plateau	2,002	26	31.4	83	54-121
Pinkerton et al., 2004	Uranium mills, Colorado Plateau	1,485	78 (includes bronchi, lungs, trachea)	68.93	113	89-141
Polednak and Frome, 1981	Y-12 uranium-processing plant, Oak Ridge, TN	18,869	324	296.47	109 122 (corrected+)	97-122 110-136 (corrected)
Checkoway et al., 1988	Y-12 uranium-fabrication plant, Oak Ridge, TN	6,781	89	65.4 (calculated)	136 (comparison with US WM)	109-167
Frome et al., 1990	Y-12, K-25 uranium-enrichment plant, research laboratory, Oak Ridge, TN	28,008	850	667.99	127	$p < 0.01$
Loomis and Wolf, 1996	Y-12 uranium-fabrication plant, Oak Ridge, TN	10,597	202 194 (WM)	172.6 (calculated) 161.7 (calculated)	117 (all workers) 120 (WM)	101-134 104-138

Frome et al., 1997	All four plants, Oak Ridge, TN (wm)	106,020	1,849		118	NS (article states increased because of high value of FTR)
Richardson and Wing, 2006	Y-12 uranium-fabrication plant, Oak Ridge, TN	3,864	40 (highest internal dose, 100+ mSv)		RR, 1.4	0.65-3.01
Hadjimichael et al., 1983	Nuclear-fuel fabricating plant, CT (men)	2,613	14	14.7	95	52-160
Stayner et al., 1985	Phosphate-fertilizer production plant, FL	3,199	10	8.85	113	61-192
Dupree et al., 1987	Uranium-processing plant, Buffalo, NY	995	21	21.7	97	60-148
Brown and Bloom, 1987	Uranium-enrichment plant, OH	5,773	48	54.6	88	65-117
Cragle et al., 1988	Nuclear-fuels production plant, Savannah, SC	9,860	53 (hourly) 22 (salaried) 8 (both)	62.04 (hourly) 33.01 (salaried) 5.27 (both)	85 (hourly) 67 (salaried) 152 (both)	NS
Beral et al., 1988	Atomic Weapons Establishment, UK (employees with radiation records)	22,552	85	132.86	64	$p < 0.01$
Ritz, 1999	Uranium-processing plant, OH	4,014	112	111.03	101	83-121
Dupree-Ellis et al., 2000	Uranium-processing plant, St Louis, MO	2,514	98	96.08 (calculated)	102	83-124

Continued

TABLE 8-1 Continued

Study	Cohort/Study Site	Population	No. Observed Cases/Deaths	No. Expected Cases/Deaths	SMR	95% CI
Ritz et al., 2000	Rocketdyne/Atomics International plant, CA	2,297	46	56.95	81	59-108
Boice et al., 2006	Rocketdyne/Atomics International plant, CA	5,801	151	NR	89	76-105
Dupree et al., 1995	Case-control study of lung cancer in four uranium-processing operations	787 lung-cancer cases	NA	NA	(OR, 2.0 for exposed to 25+ cGy)	0.20-20
McGeoghegan and Binks, 2000a	Capenhurst uranium-enrichment plant, Cheshire, UK (radiation workers)	12,540	67 deaths (50 y) 49 cases (20 y)	74.90 deaths 58.13 cases	89 (SIR, 84)	NS NS
McGeoghegan and Binks, 2000b	Springfields uranium-production plant, Lancashire, UK (radiation workers)	19,454	360 deaths (50 y) 225 cases (20 y)	421.93 deaths 301.37 cases	85 (SIR, 75)	$p < 0.01$ $p < 0.001$
Studies of Depleted-Uranium–Exposed Persons						
Macfarlane et al., 2003	Gulf War veterans, UK	51,721	14 cases (includes bronchi, lungs, trachea)	18 cases	(IRR, 0.76) (IRR, 0.41 adj++)	0.38-1.54 0.10-1.73
Gustavsson et al., 2004	Swedish UN services personnel in Balkans	9,188	1 case (male military service)	0.8	(SIR, 125)	NS
Storm et al., 2006	Danish veterans deployed to Balkans	14,012	2 cases (male)	NR	(SIR, 40)	0-140

Studies of Environmental Exposures to Uranium						
Boice et al., 2003a	Residents of municipalities near two former nuclear-material processing plants, PA	16,772	74 cases (includes trachea, bronchi, lungs, pleura)	84.4 cases	(SIR, 88)	69-110
Boice et al., 2003b	County residents near two former nuclear-material processing plants, PA	443,799	8,064 deaths	NR	(RR, 0.95)	0.93-0.98 $p < 0.05$
Boice et al., 2003c	Residents near former uranium milling and mining site, TX	12,455	224 deaths	NR	(RR, 1.08)	0.90-1.30

NOTE: FTR = Freeman-Tukey residuals, IRR = incidence rate ratio, NR = not reported, NS = not significant, OR = odds ratio, RR = relative risk, SIR = standardized incidence ratio, SMR = standardized mortality ratio, WM = white men, + = corrected for incomplete ascertainment, ++ = adjusted for smoking and alcohol consumption.

TABLE 8-2 Lung Cancer in the Oak Ridge, Tennessee, Cohort

Study	SMR or RR		95% CI or P Value	
Polednak and Frome, 1981	122		110-136	
Checkoway et al., 1988	136		109-167	
Frome et al., 1990	127		$p < 0.05$	
Loomis and Wolf, 1996	117		101-134	
Richardson and Wing, 2006	External dose (mSv)		External dose (mSv)	
	<10	Referent	<10	Referent
	10-49.9	0.92	10-49.9	0.58-1.46
	50+	1.33	50+	0.56-3.18
	Internal dose (mSv)		Internal dose (mSv)	
	<10	Referent	<10	Referent
	10-49.9	1.52	10-49.9	0.74-3.13
	50-99.9	1.20	50-99.9	0.54-2.67
	100+	1.40	100+	0.65-3.01

NOTE: CI = confidence interval, RR = rate ratio, SMR = standardized mortality ratio.

TABLE 8-3 Leukemia

Study	Cohort/Study Site	Population	No. Observed Cases/Deaths	No. Expected Cases/Deaths	SMR	95% CI
Studies of Uranium Processors						
Archer et al., 1973	Uranium mills, Colorado Plateau	622	1	0.79	127	NS
Waxweiler et al., 1983	Uranium mills, Colorado Plateau	2,002	0	4.5	0	NS
Pinkerton et al., 2004	Uranium mills, Colorado Plateau	1,485	5	7.62	66	21-153
Polednak and Frome, 1981	Y-12 uranium-processing plant, Oak Ridge, TN	18,869	40	43.57	92 102 (corrected+)	66-125 74-137
Checkoway et al., 1988	Y-12 uranium-fabrication plant, Oak Ridge, TN	6,781	4	NR	50 (comparison with US white men)	14-128
Frome et al., 1990	Y-12, K-25 uranium-enrichment plant, research laboratory, Oak Ridge, TN	28,008	92	81.17	113	NS
Loomis and Wolf, 1996	Y-12 uranium- fabrication plant, Oak Ridge, TN	10,597	11	NR	60	3-107
Frome et al., 1997	All four plants, Oak Ridge, TN (white men)	106,020	180	NR	98	NS

Continued

TABLE 8-3 Continued

Study	Cohort/Study Site	Population	No. Observed Cases/Deaths	No. Expected Cases/Deaths	SMR	95% CI
Hadjimichael et al., 1983	Nuclear-fuel fabricating plant, CT (men)	2,613	2	1.8	113	13-409
Beral et al., 1988	Atomic Weapons Establishment, UK (employees with radiation records)	22,552	4	9.16	44	NS
Cragle et al., 1988	Nuclear-fuels production plant, Savannah, SC	9,860	13 (hourly) 4 (salaried) 1 (both)	7.95 (hourly) 3.80 (salaried) 0.61 (both)	163 (hourly) 105 (salaried) 164 (both, calculated)	NS NS NS
Ritz, 1999	Uranium-processing plant, OH	4,014	13	11.21	116	62-198
Dupree-Ellis et al., 2000	Uranium-processing plant, St Louis, MO	2,514	11	NR	111	57-189
Ritz et al., 2000	Rocketdyne/Atomics International plant, CA	2,297	8	5.47	146	63-288
Boice et al., 2006	Rocketdyne/Atomics International plant, CA	5,801	25 leukemia, aleukemia deaths	NR	133	86-197
McGeoghegan and Binks, 2000a	Capenhurst uranium-enrichment plant, Cheshire, UK (radiation workers)	12,540	4 deaths (50 y) 4 cases (20 y)	5.76 deaths 5.41 cases	69 (SIR, 74)	NS NS

McGeoghegan and Binks, 2000b	Springfields uranium-production plant, Lancashire, UK (radiation workers)	19,454	32 deaths (50 y) 22 cases (20 y)	32.07 deaths 27.75 cases	100 (SIR, 79)	NS NS
Studies of Depleted-Uranium–Exposed Persons						
Gustavsson, 2004	Swedish UN services personnel in Balkans	9,188	1 chronic myeloid leukemia case	0.3 case	(SIR, 333)	NS
Storm et al., 2006	Danish veterans deployed to Balkans	14,012	4 cases	NR	(SIR, 140)	40-350
Studies of Environmental Exposures to Uranium						
Boice et al., 2003a	Residents of municipalities near two former nuclear-material processing plants, PA	16,772	18 cases	12.4 cases	145	86-230
Boice et al., 2003b	County residents near two former nuclear-material processing plants, PA	443,799	1,529 deaths	NR	(RR, 0.91)	0.86-0.97
Boice et al., 2003c	Residents near former uranium milling and mining site, TX	12,455	59 deaths	NR	(RR, 1.15)	0.9-1.6
Auvinen et al., 2002	Drinking well-water study, Finland	35 leukemia cases, 274 controls	NA	NA	(HR, 0.89) (HR, 0.91 per Bq/L for uranium)	0.38-2.11 0.73-1.13

NOTE: CI = confidence interval, HR = hazard ratio, NA = not applicable (case-control study), NR = not reported, NS = not significant, RR = relative risk, SIR = standardized incidence ratio, SMR = standardized mortality ratio, + = corrected for incomplete ascertainment.

TABLE 8-4 Hodgkin Lymphoma

Study	Cohort/Study Site	Population	No. Observed Cases/Deaths	No. Expected Cases/Deaths	SMR	95% CI
Studies of Uranium Processors						
Waxweiler et al., 1983	Uranium mills, Colorado Plateau	2,002	3	1.3	231	48-675
Pinkerton et al., 2004	Uranium mills, Colorado Plateau	1,485	4	1.21	330	90-843
Polednak and Frome, 1981	Y-12 uranium-processing plant, Oak Ridge, TN	18,869	9	16.38	55	NS
Checkoway, 1988	Y-12 uranium-fabrication plant, Oak Ridge, TN	6,781	3	NR	87 (comparison with US white men)	18-254
Frome et al., 1990	Y-12, K-25 uranium-enrichment plant, research labatory, Oak Ridge, TN	28,008	18	23	78	NS
Loomis and Wolf, 1996	Y-12 uranium-fabrication plant, Oak Ridge, TN	10,597	3	NR	62	13-183
Frome et al., 1997	All four plants, Oak Ridge, TN (white men)	106,020	40	NR	77	NS
Beral et al., 1988	Atomic Weapons Establishment, UK (employees with radiation records)	22,552	2	3.55	56	NS
Ritz, 1999	Uranium-processing plant, OH	4,014	6	2.95	204	74-443
Dupree-Ellis et al., 2000	Uranium-processing plant, St Louis, MO	2,514	2	NR	92	15-283

McGeoghegan and Binks, 2000a	Capenhurst uranium-enrichment plant, Cheshire, UK (radiation workers)	12,540	2 deaths (50 y) 1 case (20 y)	1.13 deaths 1.54 cases	177 (SIR, 65)	NS NS
McGeoghegan and Binks, 2000b	Springfields uranium-production plant, Lancashire, UK (radiation workers)	19,454	9 deaths (50 y) 10 cases (20 y)	7.23 deaths 7.22 cases	124 (SIR, 139)	NS NS
Boice et al., 2006	Rocketdyne/Atomics International plant, CA	5,801	5	NR	199	65-463
Studies of Depleted-Uranium–Exposed Persons						
Gustavsson et al., 2004	Swedish UN services personnel in Balkans	9,188	2 cases	1.1 cases	(SIR, 190)	20-670
Nuccetelli et al., 2005	Italian veterans deployed to Balkans	About 40,000	NR	NR	(SIR, 236)	122-436
Storm et al., 2006	Danish veterans deployed to Balkans	14,012	3 cases	NR	(SIR, 100)	20-290
Studies of Environmental Exposure to Uranium						
Boice et al., 2003a	Residents of municipalities near two former nuclear-material processing plants, PA	16,772	1 case	2.7 cases	(SIR, 37)	0-205
Boice et al., 2003b	County residents near two former nuclear-material processing plants, PA	443,799	290 deaths	NR	(RR, 0.97)	0.85-1.12
Boice et al., 2003c	Residents near former uranium milling and mining site, TX	12,455	12 deaths	NR	(RR, 1.79)	0.9-3.6

NOTE: CI = confidence interval, NR = not reported, NS = not significant, RR = relative risk, SIR = standardized incidence ratio, SMR = standardized mortality ratio.

TABLE 8-5 Non-Hodgkin Lymphoma and Other Lymphatic Cancers

Study	Cohort/Study Site	Population	No. Observed Cases/Deaths	No. Expected Cases/Deaths	SMR	95% CI
Studies of Uranium Processors						
Archer et al., 1973	Uranium mills, Colorado Plateau	662	4 lymphatic, hematopoietic cancers	1.02	392	$p < 0.05$
Waxweiler et al., 1983	Uranium mills, Colorado Plateau	2,002	4 lymphatic cancers	4.4	91	NS
Pinkerton et al., 2004	Uranium mills, Colorado Plateau	1,485	12 lymphatic cancers	9.86	122	NS
Polednak and Frome, 1981	Y-12 uranium-processing plant, Oak Ridge, TN	18,869	17 lymphosarcoma, reticulosarcoma	25.39	67	NS
			11 other lymphatic tissue	19.27	57	NS
Checkoway et al., 1988	Y-12 uranium-fabrication plant, Oak Ridge, TN	6,781	3 lymphosarcoma 9 other lymphatic cancers	NR NR	62 186	13-181 85-353
Frome et al., 1990	Y-12, K-25 uranium-enrichment plant, research laboratory, Oak Ridge, TN	28,008	39 lymphosarcoma, reticulosarcoma 40 other lymphatic cancers	45.80 48.23	85 83	NS NS
Loomis and Wolf, 1996	Y-12 uranium-fabrication plant, Oak Ridge, TN	10,597	4 lymphosarcoma, reticulosarcoma	NR	50	14-129
			22 other lymphatic cancers	NR	132	82-199

Frome et al., 1997	All four plants, Oak Ridge, TN (white men)	106,020	82 lymphosarcoma, reticulosarcoma	NR	91	NS
			105 other lymphatic cancers	NR	84	NS
Hadjimichael et al., 1983	Nuclear-fuel fabricating plant, CT (men)	2,613	2 lymphatic, hematopoietic cancers	3.1	65	7-234
Stayner et al., 1985	Phosphate-fertilizer production plant, FL	3,199	2 lymphatic, hematopoietic cancers	3.78	53	9-167
Brown and Bloom et al., 1987	Uranium-enrichment plant, OH	5,773	23 cancer of lymphatic, hematopoietic system (may include leukemia)	15.8	146	NS
Cragle et al., 1988	Nuclear-fuels production plant, Savannah, SC	9,860	Lymphatic cancers: 6 (hourly) 4 (salaried) 4 (both)	 9.5 (hourly) 4.73 (salaried) 0.76 (both)	 95 (hourly) 85 (salaried) 526 (both)	 NS NS NS
Beral et al., 1988	Atomic Weapons Establishment, UK (employees with radiation records)	22,552	3 NHL	6.17	49	NS
			2 multiple myeloma	3.55	56	NS
Ritz, 1999	Uranium-processing plant, OH	4,014	8 lymphosarcoma, and reticulosarcoma	4.79	167	72-329
			10 lymphatic tissue	9.94	101	48-185
Dupree-Ellis et al., 2000	Uranium-processing plant, St Louis, MO	2,514	1 lymphosarcoma	NR	28	1-156
			5 multiple myeloma	NR	130	42-303
			9 other lymphoid tissue	NR	96	43-186
Ritz et al., 2000	Rocketdyne/Atomics International plant, CA	2,297	4 lymphatic cancers	8.98	45	NS

Continued

TABLE 8-5 Continued

Study	Cohort/Study Site	Population	No. Observed Cases/Deaths	No. Expected Cases/Deaths	SMR	95% CI
Boice et al., 2006	Rocketdyne/Atomics International plant, CA	5,801	19	NR	98	59-152
McGeoghegan and Binks, 2000a	Capenhurst uranium enrichment plant, Cheshire, UK (radiation workers)	12,540	NHL:			
			5 deaths (50 y)	4.58 deaths	109	NS
			3 cases (20 y)	5.21 cases	(SIR, 58)	NS
			Multiple myeloma:	2.61 deaths	115	
			3 deaths (50 y)	2.41 cases	(SIR, 83)	NS
			2 cases (20 y)			NS
McGeoghegan and Binks, 2000b	Springfields uranium-production plant, Lancashire, UK (radiation workers)	19,454	NHL:			
			15 deaths (50 y)	23.78 deaths	63	NS
			20 cases (20 y)	25.39 cases	(SIR, 79)	NS
			Multiple myeloma:			
			11 deaths (50 y)	13.83 deaths	80	NS
			10 cases (20 y)	12.36 cases	(SIR, 81)	NS
Studies of Depleted-Uranium–Exposed Persons						
Gustavsson et al., 2004	Swedish UN services personnel in Balkans	9,188	1 case NHL	1.2 cases	(SIR, 83 calculated)	NS
Storm et al., 2006	Danish veterans deployed to Balkans	14,012	3 cases NHL	NR	(SIR, 80)	20-230
			1 case myeloma	NR	(SIR, 190)	0.0-1060

	former nuclear-material processing plants, PA					
			11 cases multiple myeloma	5.8	(SIR, 191)	95-342
Boice et al., 2003b	County residents near two former nuclear-material processing plants, PA	443,799	1,329 deaths NHL	NR	(RR, 1.06)	0.99-1.13
			561 deaths multiple myeloma	NR	(RR, 0.98)	0.89-1.09
Boice et al., 2003c	Residents near former uranium milling and mining site, TX	12,455	38 deaths NHL	NR	(RR, 1.00)	0.7-1.4
			22 deaths multiple myeloma	NR	(RR, 1.37)	0.8-2.3

NOTE: CI = confidence interval, NHL = non-Hodgkin lymphoma, NR = not reported, NS = not significant, RR = relative risk (computed as ratio of SMR of study population to that of control population), SIR = standardized incidence ratio, SMR = stanardized mortality ratio.

TABLE 8-6 Bone Cancer

Study	Cohort/Study Site	Population	No. Observed Cases/Deaths	No. Expected Cases/Deaths	SMR	95% CI
Studies of Uranium Processors						
Polednak and Frome, 1981	Y-12 uranium-processing plant, Oak Ridge, TN	18,869	6	6.68	90 100 (corrected+)	33-196 40-206
Frome et al., 1990	Y-12, K-25 uranium-enrichment plant, research laboratory, Oak Ridge, TN	28,008	11	10.35	106	NS
Loomis and Wolf, 1996	Y-12 uranium-fabrication plant, Oak Ridge, TN	10,597	1	NR	62	1-345
Frome et al., 1997	All four plants, Oak Ridge, TN (white men)	106,020	25	NR	119	NS
Hadjimichael et al., 1983	Nuclear-fuel fabricating plant, CT (men)	2,613	1	0.5	206	3-1140
Cragle et al., 1988	Nuclear-fuels production plant, Savannah, SC	9,860	0 (hourly) 1 (salaried) 0 (both)	1.03 (hourly) 0.48 (salaried) 0.8 (both)	0 (hourly) 208 (salaried, calculated) 0 (both)	NS NS
Beral et al., 1988	Atomic Weapons Establishment, UK (employees with radiation records)	22,552	1	1.35	74	NS
Ritz, 1999	Uranium-processing plant, OH	4,014	0	0.99	0	0-370
Dupree-Ellis et al., 2000	Uranium-processing plant, St Louis, MO	2,514	1	NR	120	7-526

McGeoghegan and Binks, 2000a	Capenhurst uranium-enrichment plant, Cheshire, UK (radiation workers)	12,540	0 deaths (50 y) 0 cases (20 y)	0.46 death 0.39 case	0 0	NS NS
McGeoghegan and Binks, 2000b	Springfields uranium-production plant, Lancashire, UK (radiation workers)	19,454	2 deaths (50 y) 0 cases (20 y)	2.98 deaths 1.92 cases	67 (SIR, 0)	NS
Boice et al., 2006	Rocketdyne/Atomics International plant, CA	5,801	0	1	0	0-352
Study of Depleted-Uranium–Exposed Persons						
Storm et al., 2006	Danish veterans deployed to Balkans	14,012	4 cases[a]	NR	(SIR, 600)	160-1530
Studies of Environmental Exposures to Uranium						
Boice et al., 2003b	County residents near two former nuclear-material processing plants, PA	443,799	168 deaths	NR	(RR, 1.01)	0.84-1.21
Boice et al., 2003c	Residents near former uranium milling and mining site, TX	12,455	11 deaths	NR	(RR, 1.35)	0.7-2.8

NOTE: CI = confidence interval, NR = not reported, NS = not significant, RR = relative risk (computed as ratio of SMR in study population to that in control population), SIR = standardized incidence ratio, SMR = standardized mortality ratio, + = corrected for incomplete ascertainment.

[a]Result is significant. However, three cases occurred in first year after deployment; omitting first year, SIR is 170 (95% CI, 0-1,010).

TABLE 8-7 Renal Cancer

Study	Cohort/Study Site	Population	No. Observed Cases/Deaths	No. Expected Cases/Deaths	SMR	95% CI
Studies of Uranium Processors						
Waxweiler et al., 1983	Uranium mills, Colorado Plateau	2,002	3	2.7	112	23-325
Pinkerton et al., 2004	Uranium mills, Colorado Plateau	1,485	4	4.96	81	22-206
Polednak and Frome, 1981	Y-12 uranium-processing plant, Oak Ridge, TN	18,869	20	26.54	75	NS
Checkoway et al., 1988	Y-12 uranium-fabrication plant, Oak Ridge, TN	6,781	6	NR	122 (comparison with US white men)	45-266
Frome et al., 1990	Y-12, K-25 uranium-enrichment plant, research laboratory, Oak Ridge, TN	28,008	44	52.63	84	NS
Loomis and Wolf, 1996	Y-12 uranium-fabrication plant, Oak Ridge, TN	10,597	16	NR	130	74-211
Frome et al., 1997	All four plants, Oak Ridge, TN (white men)	106,020	109	NR	92	NS
Cragle et al., 1988	Nuclear-fuels production plant, Savannah, SC	9,860	2 (hourly) 1 (salaried) 0 (both)	5.01 (hourly) 2.56 (salaried) 0.41 (both)	40 (hourly) 39 (salaried) 0 (both)	NS NS NS
Beral et al., 1988	Atomic Weapons Establishment, UK (employees with radiation records)	22,552	11	5.84	188	NS
Ritz, 1999	Uranium-processing plant, OH	4,014	5	7.89	63	20-146

Dupree-Ellis et al., 2000	Uranium-processing plant, St Louis, MO	2,514	8	NR	117	54-218
Ritz et al., 2000	Rocketdyne/Atomics International plant, CA	2,297	5	3.97	126	41-294
Boice et al., 2006	Rocketdyne/Atomics International plant, CA	5,801	12	NR	94	49-164
McGeoghegan and Binks, 2000a	Capenhurst uranium-enrichment plant, Cheshire, UK (radiation workers)	12,540	2 deaths (50 y) 2 cases (20 y)	4.08 deaths 4.47 cases	49 (SIR, 45)	NS NS
McGeoghegan and Binks, 2000b	Springfields uranium-production plant, Lancashire, UK (radiation workers)	19,454	13 deaths (50 y) 14 cases (20 y)	21.65 deaths 22.31 cases	60 (SIR, 63)	NS NS
Study of Depleted-Uranium–Exposed Persons						
Storm et al., 2006	Danish veterans deployed to Balkans	14,012	2 cases	NR	(SIR, 110)	10-410
Studies of Environmental Exposures to Uranium						
Boice et al., 2003a	Residents of municipalities near two former nuclear-material processing plants, PA	16,772	14 cases	13.3 cases	(SIR, 105)	58-177
Boice et al., 2003b	County residents near two former nuclear-material processing plants, PA	443,799	784 deaths	NR	(RR, 1.02)	0.94-1.12
Boice et al., 2003c	Residents near former uranium milling and mining site, TX	12,455	19	NR	(RR, 0.58)	0.4-1.0 $p < 0.05$
Kurttio et al., 2006b	Drinking well-water study, Finland	51 renal-cancer cases, 274 controls	NA	NA	(HR, 0.74) (HR, 0.92 per log [1 Bq/L] for uranium)	0.33-1.66 0.36-2.35

NOTE: CI = confidence interval, HR = hazard ratio, NA = not applicable (case-control study), NR, = not reported, NS = not significant, RR = relative risk, SIR = standardized incidence ratio, SMR = standardized mortality ratio.

TABLE 8-8 Bladder Cancer

Study	Cohort/Study Site	Population	No. Observed Cases/Deaths	No. Expected Cases/Deaths	SMR	95% CI
Studies of Uranium Processors						
Polednak and Frome, 1981	Y-12 uranium-processing plant, Oak Ridge, TN	18,869	26	32.32	80	NS
Checkoway et al., 1988	Y-12 uranium-fabrication plant, Oak Ridge, TN	6,781	3	NR	72 (comparison with US white men)	15-210
Frome et al., 1990	Y-12, K-25 uranium-enrichment plant, research laboratory, Oak Ridge, TN	28,008	54	66.22	82	NS
Loomis and Wolf, 1996	Y-12 uranium-fabrication plant, Oak Ridge, TN	10,597	8	NR	72	31-142
Frome et al., 1997	All four plants, Oak Ridge, TN (white men)	106,020	105	NR	76	NS
Hadjimichael et al., 1983	Nuclear-fuel fabricating plant, CT (men)	2,613	1	1.9	0.52	1-292
Cragle et al., 1988	Nuclear-fuels production plant, Savannah, SC	9,860	2 (hourly) 4 (salaried) 0 (both)	3.34 (hourly) 2.14 (salaried) 0.3 (both)	60 (hourly) 187 (salaried) 0 (both)	NS NS NS
Beral et al., 1988	Atomic Weapons Establishment, UK (employees with radiation records)	22,552	7	13.69	51	NS
Ritz, 1999	Uranium-processing plant, OH	4,014	8	6.95	115	50-227
Dupree-Ellis et al., 2000	Uranium-processing plant, St Louis, MO (white men)	2,514	8	NR	116	48-236

Ritz et al., 2000	Rocketdyne/Atomics International plant, CA	2,297	3	3.39	89	18-259
Boice et al., 2006	Rocketdyne/Atomics International plant, CA	5,801	8	NR	65	28-129
McGeoghegan and Binks, 2000a	Capenhurst uranium-enrichment plant, Cheshire, UK (radiation workers)	12,540	8 deaths (50 y) 14 cases (20 y)	7.69 deaths 14.57 cases	104 (SIR, 96)	NS NS
McGeoghegan and Binks, 2000b	Springfields uranium-production plant, Lancashire, UK (radiation workers)	19,454	40 deaths (50 y) 57cases (20 y)	43.66 deaths 75.15 cases	92 (SIR, 76)	NS $p < 0.05$
Studies of Depleted-Uranium–Exposed Persons						
Gustavsson et al., 2004	Swedish UN services personnel in Balkans	9,188	2 cases	0.7 case	(SIR, 290)	40-1100
Storm et al., 2006	Danish veterans deployed to Balkans	14,012	7 cases	NR	(SIR, 220)	90-450
Studies of Environmental Exposures to Uranium						
Boice et al., 2003a	Residents of municipalities near two former nuclear-material processing plants, PA	16,772	36 cases	30.2 case	(SIR, 119)	83-165
Boice et al., 2003b	County residents near two former nuclear-material processing plants, PA	443,799	1,044 deaths	NR	(RR, 0.97)	0.9-1.04
Boice et al., 2003c	Residents near former uranium milling and mining site, TX	12,455	17 deaths	NR	(RR, 0.64)	0.4-1.1
Kurttio et al., 2006b	Drinking well-water study, Finland	61 bladder-cancer cases, 274 controls	NA	NA	(HR, 0.77 per log [1Bq/L] for uranium)	0.41-1.98

NOTE: CI = confidence interval, HR = hazard ratio, NA = not applicable (case-control study), NR = not reported, NS = not significant, RR = relative risk, SIR = standardized incidence ratio, SMR = standardized mortality ratio.

TABLE 8-9 Cancers of the Central Nervous System

Study	Cohort/Study Site	Population	No. Observed Cases/Deaths	No. Expected Cases/Deaths	SMR	95% CI
Studies of Uranium Processors						
Polednak and Frome, 1981	Y-12 uranium-processing plant, Oak Ridge, TN	18,869	32	33.83	95	NS
Checkoway et al., 1988	Y-12 uranium-fabrication plant, Oak Ridge, TN	6,781	14	NR	180 (comparison with US white men)	98-302
Frome et al., 1990	Y-12, K-25 uranium-enrichment plant, research laboratory, Oak Ridge, TN	28,008	69	59.57	116	NS
Loomis and Wolf, 1996	Y-12 uranium-fabrication plant, Oak Ridge, TN	10,597	20	NR	129	79-200
Frome et al., 1997	All four plants, Oak Ridge, TN (white men)	106,020	151 (brain only)	NR	109	NS
Hadjimichael et al., 1983	Nuclear-fuel fabricating plant, CT (men)	2,613	4	1.7	240	65-615
Cragle et al., 1988	Nuclear-fuels production plant, Savannah, SC	9,860	2 (hourly) 4 (salaried) 1 (both)	8.40 (hourly) 3.77 (salaried) 0.64 (both)	23 (hourly) 106 (salaried) 156 (both)	$p < 0.05$ NS NS
Beral et al., 1988	Atomic Weapons Establishment, UK (employees with radiation records)	22,552	3	9.36	32	$p < 0.05$
Ritz, 1999	Uranium-processing plant, OH	4,014	12	9.66	124	64-217
Dupree-Ellis et al., 2000	Uranium-processing plant, St Louis, MO	2,514	12	NR	157	84-264

Ritz et al., 2000	Rocketdyne/Atomics International plant, CA	2,297	6	4.6	131	48-284
Boice et al., 2006	Rocketdyne/Atomics International plant, CA	5,801	17	NR	115	67-183
McGeoghegan and Binks, 2000a	Capenhurst uranium-enrichment plant, Cheshire, UK (radiation workers)	12,540	7 deaths (50 y) 4 cases (20 y)	5.04 deaths 3.89 cases	139 (SIR, 103)	NS NS
McGeoghegan and Binks, 2000b	Springfields uranium-production plant, Lancashire, UK (radiation workers)	19,454	18 deaths (50 y) 12 cases (20 y)	27.03 deaths 18.76 cases	67 (SIR, 64)	NS NS
Studies of Depleted-Uranium–Exposed Persons						
Macfarlane et al., 2003	Gulf War veterans, UK	51,721	21 cases	25 cases	(IRR, 0.83) (IRR, 1.08 adjusted++)	0.46-1.48 0.44-2.65
Gustavsson et al., 2004	Swedish UN services personnel in Balkans	9,188	3 cases	2.6 cases	(SIR, 120)	20-340
Storm et al., 2006	Danish veterans deployed to Balkans	14,012	9 cases	NR	(SIR, 120)	50-220
Studies of Environmental Exposures to Uranium						
Boice et al., 2003a	Residents of municipalities near two former nuclear-material processing plants, PA	16,772	3 cases	6.7 cases	(SIR, 45)	9-130
Boice et al., 2003b	County residents near two former nuclear-material processing plants, PA	443,799	779 deaths	NR	(RR, 0.96)	0.88-1.04
Boice et al., 2003c	Residents near former uranium milling and mining site, TX	12,455	24 deaths	NR	(RR, 0.92)	0.6-1.4

NOTE: CI = confidence interval, IRR = incidence rate ratio, NR = not reported, NS = not significant, RR = relative risk, SIR = standardized incidence ratio, SMR = standardized mortality ratio, ++ = adjusted for smoking and alcohol consumption.

TABLE 8-10 Stomach Cancer

Study	Cohort/Study Site	Population	No. Observed Cases/Deaths	No. Expected Cases/Deaths	SMR	95% CI
Studies of Uranium Processors						
Waxweiler et al., 1983	Uranium mills, Colorado Plateau	2,002	3	7.5	40	8-117
Polednak and Frome, 1981	Y-12 uranium-processing plant, Oak Ridge, TN	18,869	53	72.33	73	NS
Checkoway et al., 1988	Y-12 uranium-fabrication plant, Oak Ridge, TN	6,781	5	NR	57 (comparison with US white men)	19-133
Frome et al., 1990	Y-12, K-25 uranium-enrichment plant, research laboratory, Oak Ridge, TN	28,008	93	119.92	78	NS
Loomis and Wolf, 1996	Y-12 uranium-fabrication plant, Oak Ridge, TN	10,597	12	NR	64	33-112
Frome et al., 1997	All four plants, Oak Ridge, TN (white men)	106,020	176	NR	73	NS
Dupree et al., 1987	Uranium-processing plant, Buffalo, NY	995	7	4.2	165	66-339
Brown and Bloom, 1987	Uranium-enrichment plant, OH	5,773	10	5.9	169	NS
Cragle et al., 1988	Nuclear-fuels production plant, Savannah, SC	9,860	5 (hourly) 2 (salaried) 0 (both)	7.38 (hourly) 4.15 (salaried) 0.64 (both)	68 (hourly) 48 (salaried) 0 (both)	NS NS NS
Beral et al., 1988	Atomic Weapons Establishment, UK (employees with radiation records)	22,552	24	35.78	67	$p < 0.05$
Ritz, 1999	Uranium-processing plant, OH	4,014	15	11.18	134	75-221
Dupree-Ellis et al., 2000	Uranium-processing plant, St Louis, MO	2,514	4	NR	38	12-89

Ritz et al., 2000	Rocketdyne/Atomics International plant, CA	2,297	6	5.07	118	43-257
Boice et al., 2006	Rocketdyne/Atomics International plant, CA	5,801	21	NR	117	73-179
McGeoghegan and Binks, 2000a	Capenhurst uranium-enrichment plant, Cheshire, UK (radiation workers)	12,540	15 deaths (50 y) 13 cases (20 y)	16.61 deaths 13.94 cases	90 (SIR, 93)	NS NS
McGeoghegan and Binks, 2000b	Springfields uranium-production plant, Lancashire, UK (radiation workers)	19,454	92 deaths (50 y) 56 cases (20 y)	99.95 deaths 73.90 cases	92 (SIR, 76)	NS $p < 0.05$
Study of Depleted-Uranium–Exposed Persons						
Storm et al., 2006	Danish veterans deployed to Balkans	14,012	2 cases	NR	(SIR, 160)	20-560
Studies of Environmental Exposures to Uranium						
Boice et al., 2003a	Residents of municipalities near two former nuclear-material processing plants, PA	16,772	10 cases	9.5 cases	(SIR, 105)	50-193
Boice et al., 2003b	County residents near two former nuclear-material processing plants, PA	443,799	2,203 deaths	NR	(RR, 1.00)	0.95-1.06
Boice et al., 2003c	Residents near former uranium milling and mining site, TX	12,455	72 deaths	NR	(RR, 1.08)	0.8-1.4
Auvinen et al., 2005	Drinking well-water study, Finland	Case-control study: 88 stomach-cancer cases, 274 controls	NA	NA	(HR, 0.69) (HR, 0.76 per Bq/L for uranium)	0.37-1.27 0.48-1.21

NOTE: CI = confidence interval, HR = hazard ratio, NA = not applicable (case-control study), NR = not reported, NS = not significant, RR = relative risk, SIR = standardized incidence ratio, SMR = standardized mortality ratio.

TABLE 8-11 Prostatic Cancer

Study	Cohort/Study Site	Population	No. Observed Cases/Deaths	No. Expected Cases/Deaths	SMR	95% CI
Studies of Uranium Processors						
Polednak and Frome, 1981	Uranium mills, Colorado Plateau	1,485	49	60.71	81	NS
Waxweiler et al., 1983	Uranium mills, Colorado Plateau	2,002	6	8.5	71	26-154
Checkoway et al., 1988	Y-12 uranium-fabrication plant, Oak Ridge, TN	6,781	7	NR	92 (comparison with US white men)	37-190
Frome et al., 1990	Y-12, K-25 uranium-enrichment plant, research laboratory, Oak Ridge, TN	28,008	150	141.96	106	NS
Loomis and Wolf, 1996	Y-12 uranium-fabrication plant, Oak Ridge, TN	10,597	36	NR	131	91-181
Frome et al., 1997	All four plants, Oak Ridge, TN (white men)	106,020	319	NR	101	NS
Cragle et al., 1988	Nuclear-fuels production plant, Savannah, SC	9,860	3 (hourly) 5 (salaried) 0 (both)	5.00 (hourly) 3.69 (salaried) 0.47 (both)	60 (hourly) 135 (salaried) 0 (both)	NS NS NS
Beral et al., 1988	Atomic Weapons Establishment, UK (employees with radiation records)	22,552	20	14.36	139	NS
Ritz, 1999	Uranium-processing plant, OH	4,014	25	17.42	144	93-212
Dupree-Ellis et al., 2000	Uranium-processing plant, St Louis, MO	2,514	23	NR	115	74-170
Ritz et al., 2000	Rocketdyne/Atomics International plant, CA	2,297	7	9.59	73	29-150

Boice et al., 2006	Rocketdyne/Atomics International plant, CA	5,801	37	NR	93	66-129
McGeoghegan and Binks, 2000a	Capenhurst uranium-enrichment plant, Cheshire, UK (radiation workers)	12,540	11 deaths (50 y) 9 cases (20 y)	13.91 deaths 16.72 cases	79 (SIR, 54)	NS NS
McGeoghegan and Binks, 2000b	Sprinfields uranium-production plant, Lancashire, UK (radiation workers)	19,454	68 deaths (50 y) 69 cases (20 y)	76.71 deaths 89.79 cases	89 (SIR, 77)	NS $p < 0.05$
Studies of Depleted-Uranium–Exposed Persons						
Macfarlane et al., 2003	Swedish UN services personnel in Balkans	51,721	7 cases	6 cases	(IRR, 1.15) (IRR, 1.03 adj++)	0.39-3.41 0.23-4.62
Storm et al., 2006	Danish veterans deployed to Balkans	14,012	1 cases	NR	(SIR, 60)	0-330
Studies of Environmental Exposures to Uranium						
Boice et al., 2003a	Residents of municipalities near two former nuclear-material processing plants, PA	16,772	81 cases	83.2 cases	(SIR, 97)	77-121
Boice et al., 2003b	County residents near two former nuclear-material processing plants, PA	443,799	2,181 deaths	NR	(RR, 0.95)	0.9-1.0
Boice et al., 2003c	Residents near former uranium milling and mining site, TX	12,455	76 deaths	NR	(RR, 0.95)	0.7-1.2

NOTE: CI = confidence interval, IRR = incidence rate ratio, NR = not reported, NS = not significant, RR = relative risk, SIR = standardized incidence ratio, SMR = standardized mortality ratio, ++ = adjusted for smoking and alcohol consumption.

TABLE 8-12 Testicular Cancer

Study	Cohort/Study Site	Population	No. Observed Cases/Deaths	No. Expected Cases/Deaths	SMR	95% CI
Studies of Uranium Processors						
Dupree-Ellis et al., 2000	Uranium-processing plant, St Louis, MO	2,514	1	NR	93	5-408
Ritz, 1999	Uranium-processing plant, OH	4,014	1	1.46	67	1-374
Frome et al., 1997	All four plants, Oak Ridge, TN (white men)	106,020	18	NR	72	NS
Beral et al., 1988	Atomic Weapons Establishment, UK	22,552	1	1.72	58	NS
Pinkerton et al., 2004	Uranium mills, Colorado Plateau	1,485	15	19.67	76	43-126
Polednak and Frome, 1981	Y-12 uranium-processing plant, Oak Ridge, TN	18,869	4	7.29	55	NS
Frome et al., 1990	Y-12, K-25 uranium-enrichment plant, research laboratory, Oak Ridge, TN	28,008	7	9.61	73	NS
Loomis and Wolf, 1996	Y-12 uranium-fabrication plant, Oak Ridge, TN	10,597	0	NR	0	0-159

McGeoghegan and Binks, 2000a	Capenhurst uranium-enrichment plant, Cheshire, UK (radiation workers)	12,540	0 deaths (50 yrs) 2 cases (20 yrs)	0.54 death 2.08 cases	0 (SIR, 96)	NS NS
McGeoghegan and Binks, 2000b	Springfields uranium-production plant, Lancashire, UK (radiation workers)	19,454	2 deaths (50 yrs) 8 cases (20 yrs)	3.29 deaths 8.70 cases	61 (SIR, 92)	NS NS
Boice et al., 2006	Rocketdyne/Atomics International plant, CA	5,801	1	NR	69	2-382
Studies of Depleted-Uranium–Exposed Persons						
Macfarlane et al., 2003	Gulf War veterans, UK	51,721	39 cases	46 cases	(IRR, 0.83) (IRR, 1.17 adj++)	0.54-1.28 0.61-2.23
Gustavsson et al., 2004	Swedish UN services personnel in Balkans	9,188	8 cases	4.3 cases	(SIR, 190)	80-370
Storm et al., 2006	Danish veterans deployed to Balkans	14,012	24 cases	NR	(SIR, 120)	80-180
Study of Environmental Exposures to Uranium						
Boice et al., 2003a	Residents of municipalities near two former nuclear-material processing plants, PA	16,772	2 cases	2.11 cases	(SIR, 95)	11-342

NOTE: CI = confidence interval, IRR = incidence rate ratio, NR = not reported, NS = not significant, SIR = standardized incidence ratio, SMR = standardized mortality ratio, ++ = adjusted for smoking and alcohol consumption.

TABLE 8-13 Mortality from Nonmalignant Renal Disease

Study	Cohort/Study Site	Population	No. Observed Deaths	No. Expected Deaths	SMR (95% CI)	Disease Classification
Waxweiler et al., 1983	Uranium mills, Colorado Plateau	2,002	6	3.6	167 (60-353)	ICD-7 592-594
Pinkerton et al., 2004	Uranium mills, Colorado Plateau	1,484	8	5.91	135 (58-267)	ICD-9 582-583, 585-587
Ritz, 1999	Uranium-processing plant, Fernald, OH	4,014	3	14.25	21 (4-129)	ICDA-8 580-629
Checkoway et al., 1988	Y-12 uranium-materials fabrication plant, Oak Ridge, TN	6,781	8	11.1	72 (31-142)	ICD-8 580-629
Frome et al., 1990	Y-12, K-25 uranium-enrichment facilities, research laboratory, Oak Ridge, TN	28,008	52	52.65	99 (71-126)[a]	ICDA-8 582
Polednak and Frome, 1981	Y-12 uranium-processing plant, Oak Ridge, TN	18,869	30	39.14	77 (45-109)[a]	Chronic nephritis
Frome et al., 1997	Four uranium-processing plants, Oak Ridge, TN	27,982	270	325.3[b]	83 (NS)[c]	ICDA-8 580-629
Ritz et al., 2000	Rocketdyne/Atomics International	2,297	5	6.44	78 (25-181)	ICD-8 580-629

Boice et al., 2006	Rocketdyne/Atomics International	5,801	12	NR	118 (61-206)	Nephritis and nephrosis
Dupree-Ellis et al., 2000	Mallinckrodt Chemical works plant, St. Louis, MO	2,514	6	3.19	188 (75-381)	ICD-8 582
McGeoghegan and Binks, 2000b	British Nuclear Fuels plant, Springfields site	19,454	28	48.94	57 ($p < 0.01$)	Genitourinary diseases
McGeoghegan and Binks, 2001	British Nuclear Fuels plant, Chapelcross site	2,628	5	4.63	108 (NS)	Genitourinary diseases
McGeoghegan and Binks, 2000a	British Nuclear Fuels plant, Capenhurst	12,543	7	7.13	98 (NS)	Genitourinary diseases
Cragle et al., 1988	Nuclear-fuels production facility, Savannah River Plant, SC	9,860	4	10.27	39 (10-96)[d]	ICDA-8 580-629

NOTE: CI = confidence interval, ICD = International Classification of Diseases, ICDA = International Classification of Diseases, Adapted, NR = not reported, SMR = standardized mortality ratio.
[a]CI calculated by Committee on Health Effects Associated with Exposure During the Gulf War; not stated in study (IOM, 2000).
[b]Number of expected deaths calculated by committee; not stated in study (Frome, 1997).
[c]SMR for white men only.
[d]SMR for hourly workers only.

TABLE 8-14 Nonmalignant Renal Disease—Morbidity Risk

Study	Population	Exposure	Health Outcomes or Outcome Measures	Results	Adjustments
Gulf War Veterans Depleted-Uranium Surveillance Study					
McDiarmid et al., 2000 Case series	29 exposed Gulf War veterans exposed to DU during friendly-fire incidents in February 1991, 38 unexposed veterans; examined in March-June 1997, 7 years after first exposure	Exposure to DU by friendly fire, may have inhaled or ingested airborne DU particles, experienced wound contamination by DU; assessed urinary, seminal uranium concentration	Serum creatinine, beta-2-microglobulin, retinol-binding protein, serum uric acid, urinary creatinine, urinary protein	No statistically significant differences in renal function between low- and high-exposure groups	Stratification at median into low and high result groups
McDiarmid et al., 2001 Case series	50 exposed Gulf War veterans divided into low-uranium, high-uranium groups; examined in March-July 1999, 8 years after first exposure	Exposure to DU by friendly fire, may have inhaled or ingested airborne DU particles, experienced wound contamination by DU; assessed urinary uranium concentration	Same as in McDiarmid et al., 2000	No statistically significant differences in renal function between low- and high-uranium-exposure groups Mean values in normal range	
McDiarmid et al., 2004 Case series	39 Gulf War veterans exposed to DU during friendly-fire incidents in February 1991; followup 1994-2001	Same exposure as in McDiarmid et al., 2001	Serum calcium, serum phosphate, urinary calcium, urinary phosphate, measures in McDiarmid et al., 2000	Renal-function measures (normal range): Serum creatinine (0.5-1.1 mg/dL); low-uranium group, 0.95 ± 0.03; high-uranium group, 0.85 ± 0.03; $p = 0.03$	

				Urinary retinol-binding protein (3-610 μg/g of creatinine); low-uranium group, 46.13 ± 3.46; high-uranium group, 65.68 ± 11.11; p = 0.06 Urinary total protein (0-92.8 mg/g of creatinine); low-uranium group, 54.63 ± 4.94; high-uranium group, 78.69 ± 10.52; p = 0.01
McDiarmid et al., 2006 Case series	32 Gulf War veterans exposed to DU during friendly-fire incidents; examined in April-July 2003, 12 years after first exposure	Same exposure as in McDiarmid et al., 2001	Urinary IAP, NAG, urinary microalbumin, measures in McDiarmid et al., 2004	Renal-function measures (normal range): Serum PO4 (2.7-4.5 mg/dL); low-uranium group, 3.75 ± 0.11; high-uranium group, 4.11 ± 0.12; p = 0.03 No statistically significant differences between low- and high-uranium groups for other renal measures

Continued

TABLE 8-14 Continued

Study	Population	Exposure	Health Outcomes or Outcome Measures	Results	Adjustments
McDiarmid et al., 2007 Case series	34 Gulf War veterans exposed to DU during friendly-fire incidents in 1991; examined in April-June 2005, 14 years after first exposure	Exposure to DU by friendly fire, may have inhaled or ingested airborne DU particles, experienced wound contamination by DU; assessed urinary uranium concentration, both current and cumulative exposure measures reported	Creatinine clearance, urinary glucose, measures in McDiarmid et al., 2006	Renal-function measures (normal range): Mean serum uric acid (3.4-7 mg/dL); low-cumulative-uranium group, 6.19 ± 0.26; high-cumulative-uranium group, 5.22 ± 0.46; $p = 0.03$ No statistically significant differences between high- and low-urinary-uranium exposure groups for other renal measures	
Drinking-Water and Residential Exposure					
Kurttio et al., 2002 Cross-sectional	325 people in Finland who obtain drinking water from drilled wells used an average of 13 years	Median drinking-water uranium concentration, 28 µg/L (interquartile range, 6-135 µg/L; maximum, 1,920 (µg/L) Median urinary uranium concentration, 13 ng/mmol of creatinine (2-75)	Urinary, serum calcium, phosphate, glucose, albumin, creatinine, beta-2-microglobulin as biomarkers of renal function	Statistically significant association between uranium exposure from drinking water and tubular function: calcium fractional excretion, $p < 0.05$ for all types of exposure; phosphate fractional excretion, $p = 0.03$ for urinary uranium	Uranium exposure adjusted for age, sex, and body mass index

		Median daily uranium intake, 39 μg (7-224)		No association between uranium exposure and glomerular function markers or remaining tubular markers	
Kurttio et al., 2006a (same population as Kurttio et al., 2002) Cross-sectional	95 men, 98 women in Finland who obtain drinking water from drilled wells an average of 16 years	Median drinking-water uranium concentration, 25 μg/L (interquartile range, 5-148 μg/L; maximum, 1,500 μg/L)	Various enzymes, creatinine, calcium, phosphate, glucose as indicators of renal-cell toxicity, renal dysfunction	Indicators of renal function within reference values; uranium in urine, hair, nails, drinking water not statistically significantly associated with indicators of cell toxicity, renal proximal tubular function, glomerular function	Sex, age (linear/ quadratic), body mass index, smoking, use of analgesics
		Median drinking-water uranium concentration, 25 μg/L (interquartile range, 5-148 μg/L; maximum, 1,500 μg/L)	Exposure assessment: uranium in drinking water, hair, nails, urine	Statistically significant association between cumulative uranium intake and glucose excretion ($p = 0.02$), between uranium exposure and increased blood pressure (diastolic, $p = 0.01$; systolic, $p = 0.07$)	

Continued

TABLE 8-14 Continued

Study	Population	Exposure	Health Outcomes or Outcome Measures	Results	Adjustments
Pinney et al., 2003 Cohort	8,496 people in FMMP; comparison rates NHIS (and NHANES, not listed)	Residential proximity (less than 2 miles) to Fernald uranium-processing plant in direction of groundwater runoff or possible well or cistern contamination in January 1952-December 1984	Renal disease	All renal disease: SPR, 215 (99% CI, 186-248) All bladder disease: SPR, 132 (99% CI, 111-156) Kidney stones: SPR, 398 (99% CI, 336-468) Renal infections: SPR, 71 (99% CI, 46-106) Other kidney trouble, NEC: SPR, 111 (99% CI, 68-172) Chronic nephritis: SPR, 203 (99% CI, 76-435) Bladder infections: SPR, 59 (99% CI, 40-84) Other bladder disorders: SPR, 17 (99% CI, 7-33) "Remaining kidney disorders": SPR, 196 (99% CI, 73-419)	Age, sex

				"Remaining bladder disorders": SPR, 809 (99% CI, 663-977)	
Occupational Exposure					
Boiano et al., 1989 Cross-sectional	NIOSH report on 146 (70%) of 208 eligible long-term employees at FFMPC after releases of uranium oxide from dust collectors in November-December 1984	Self-reported exposure incidents, job history, assessed urinary-uranium data	Serum: beta-2-microglobulin, retinol-binding protein, albumin, total protein, creatinine Urine: uranium, beta-2-microglobulin, retinol-binding protein, NAG, gamma-glutamyl transpeptidase, alanine aminopeptidase, creatinine, total protein, albumin	No associations between glomerular and tubular markers and measures of uranium exposure	Smoking
Shawky et al., 2002 Cross-sectional	86 processors at three sites in Egypt, 13 of whom also participated in urinary-uranium analysis	Air uranium concentration, 22.6×10^{-7} -11.1×10^{-5} Bq/cm3 Exposure, 1-80 μSv/h	Renal-function indicators: serum creatinine, urea, urinary uranium	Mean urinary uranium concentration, 17.8 μg/L; in subgroup from ore-crushing site, 29.2 μg/L Uranium excretion more than 20 times occupational-exposure decision level	

NOTE: BDI = Beck Depression Inventory, CI = confidence interval, DU = depleted uranium, FFMPC = Fernald Feed Materials Production Center, FMMP = Fernald Medical Monitoring Program, IAP = intestinal alkaline phosphatase, NAG = urinary *N*-acetyl-beta-glucominidase, NEC = not elsewhere classified, NHANES = National Health and Nutrition Examination Survey, NHIS = National Health Interview Survey, NIOSH = National Institute for Occupational Safety and Health, SPR = standardized prevalence ratio.

TABLE 8-15 Mortality from Nonmalignant Respiratory Disease

Study	Cohort/Study Site	Population	No. Observed Deaths	No. Expected Deaths	SMR (95% CI)	Disease Classification
Waxweiler et al., 1983	Uranium mills, Colorado Plateau	2,002	55	33.7	163 (123-212)	ICD-7 470-527
Pinkerton et al., 2004	Uranium mills, Colorado Plateau	1,484	100	70.16	143 (116-173)	ICD-9 460-519
		State rates	94	79.32	1.9 (0.96-1.45)	
Ritz, 1999	Uranium-processing plant, OH	4,014	53	79.78	66 (50-87)	ICDA-8 460-519
Checkoway et al., 1988	Y-12 uranium-materials fabrication plant, Oak Ridge, TN	6,781	37	48.9	76 (53-104)	ICD-8 460-519
Frome et al., 1990	Y-12, K-25 uranium-enrichment facilities, research laboratory, Oak Ridge, TN	28,008	792	634.11	125 (117-133)[a]	ICDA-8 460-519
Polednak and Frome, 1981	Y-12 uranium-processing plant, Oak Ridge, TN	18,869	340	310.11	122 (110-136)[b]	Diseases of respiratory system
Frome et al., 1997	Four uranium-processing plants, Oak Ridge, TN	27,982	1,568	1,400[c]	112 (NS)	ICDA-8 460-519

Ritz et al., 2000	Rocketdyne/Atomics International	2,297	30	40.26	75 (50-106)	ICD-8 460-519
Boice et al., 2006	Rocketdyne/Atomics International	5,801	68	NR	67 (52-84)	ICD-9 460-479, 488-519
Dupree-Ellis et al., 2000	Mallinckrodt Chemical works plant, St. Louis, MO	2,514	64	80	80 (62-101)	ICD-8 460-519
McGeoghegan and Binks, 2001	British Nuclear Fuels plant, Chapelcross site	2,628	22	45.43	48 ($p < 0.01$)[d]	Diseases of respiratory system
McGeoghegan and Binks, 2000b	British Nuclear Fuels plant, Springfields site	19,454	379	481.09	79 ($p = 0.02$)	Diseases of respiratory system
McGeoghegan and Binks, 2000a	British Nuclear Fuels plant, Capenhurst	12,543	53	75.62	70 ($p = 0.008$)	Diseases of respiratory system
Cragle et al., 1988	Nuclear-fuels production facility, Savannah River Plant, SC	9,860	17	41.02	41 (24-66)[e]	ICDA-8 460-519

NOTE: CI = confidence interval, ICD = International Classification of Diseases, ICDA = International Classification of Diseases, Adapted, NR = not reported, NS = not significant, SMR = standardized mortality ratio.

[a]Confidence interval calculated by Committee on Health Effects Associated with Exposure During the Gulf War; not stated in study (IOM, 2000).

[b]Corrected for incomplete ascertainment of deaths and for deaths of unknown cause.

[c]Number of expected deaths calculated by committee; not stated in study.

[d]SMR based on population rates for England and Wales.

[e]Listed SMR for hourly workers only.

TABLE 8-16 Nonmalignant Respiratory Disease—Morbidity Risk

Study	Population	Exposure	Outcomes	Results	Adjustments	Comments
Boiano et al., 1989 Cross-sectional	146 (70%) of 208 eligible long-term employees at FFMPC after releases of uranium oxide from dust collectors in November-December 1984	Self-reported exposure incidents, job history, assessed urinary-uranium data	Lung function, symptoms	Ratio of FEV_1 to FVC associated with job-history–derived uranium-exposure index; other spirometry results not associated; shortness of breath significantly associated with self-reported uranium-exposure incidents	Smoking	Limitations in exposure partly based on recall; crude, imprecise exposure categories (low, medium, high)
Pinney et al., 2003 Cohort	8,464 people in FMMP; comparison rates NHIS (and NHANES, not listed)	Residential proximity (less than 2 miles) to FFMPC in direction of groundwater runoff or possible well or cistern contamination in January 1952-December 1984	Self-reported symptoms of chronic bronchitis, asthma, emphysema	Asthma: SPR, 85 (99% CI, 73-98) Chronic bronchitis: SPR, 19 (99% CI, 14-24) Emphysema: SPR, 61 (99% CI, 41-86)	Age, sex	Study questionnaires not directly comparable; FMMP self-selected volunteer group

NOTE: CI = confidence interval, FEV_1 = forced expiratory volume in 1 second, FFMPC = Fernald Feed Materials Production Center, FMMP = Fernald Medical Monitoring Program, FVC = forced vital capacity, NHANES = National Health and Nutrition Examination Survey, NHIS = National Health Interview Study, SPR = standardized prevalence ratio.

TABLE 8-17 Mortality from Neurologic Disease

Study	Cohort/Study Site	Population	No. Observed Deaths	No. Expected Deaths	SMR (95% CI)	Disease Classification
Frome et al., 1990	Y-12, K-25 uranium-enrichment facilities, research laboratory, Oak Ridge, TN	28,008	76	81.76	93 (71-115)[a]	ICDA-8 320-389
Polednak and Frome, 1981	Y-12 uranium-processing plant, Oak Ridge, TN	18,869	38	49.3	77 (49-105)[a]	Diseases of nervous system
Dupree-Ellis et al., 2000	Mallinckrodt Chemical works plant, St. Louis, MO	2,514	11	13.41	82 (43-141)	ICD-8 320-389
Frome et al., 1997	Four uranium-processing plants, Oak Ridge, TN	27,982	148	211.43[b]	70 (NS)	ICDA-8 320-329
Boice et al., 2006	Rocketdyne/Atomics International	5,801	30	NR	96 (65-137)	ICD-9 320-389
McGeoghegan and Binks, 2000b	British Nuclear Fuels plant, Springfields site	19,454	40	58.25	69 ($p < 0.05$)	Diseases of nervous, sense organs
McGeoghegan and Binks, 2000a	British Nuclear Fuels plant, Capenhurst	12,543	10	10.25	98 (NS)	Diseases of nervous, sense organs
McGeoghegan and Binks, 2001	British Nuclear Fuels plant, Chapelcross site	2,628	5	7.06	71 (NS)	Diseases of nervous, sense organs
Cragle et al., 1988	Nuclear-fuels production facility, Savannah River plant, SC	9,860	8	9.92	81 (NS)[c]	ICDA-8 320-389

NOTE: CI = confidence interval, ICD = International Classification of Diseases, ICDA = International Classification of Diseases, Adapted, NR = not reported, NS = not significant, SMR = standardized mortality ratio.
[a]Confidence interval calculated by Committee on Health Effects Associated with Exposure During the Gulf War; not stated in study (IOM, 2000).
[b]Number of expected deaths calculated by committee; not stated in study.
[c]Listed SMR for hourly workers only.

TABLE 8-18 Reproductive and Developmental Effects

Study	Population	Exposure	Outcomes or Outcome Measures	Results	Adjustments
McDiarmid et al., 2000 Case series	29 exposed Gulf War veterans exposed to DU during friendly-fire incidents in February 1991, 38 unexposed veterans, examined in March-June 1997	Exposure to DU by friendly fire, may have inhaled, ingested airborne DU particles, experienced wound contamination by DU; assessed urinary and seminal uranium concentration	Neuroendocrine measures: FSH, LH, prolactin, testosterone; semen characteristics	Prolactin, 2.1-17.7 µg/g of creatinine; low urinary uranium, 1.66; high urinary uranium, 12.47; $p = 0.04$	Stratification at median into low-, high-result groups
McDiarmid et al., 2001 Case series	50 exposed Gulf War veterans divided into low-uranium and high-uranium groups, examined in March-July 1999	Exposure to DU by friendly fire, may have inhaled, ingested airborne DU particles, experienced wound contamination by DU; assessed urinary uranium concentration	Neuroendocrine measures: FSH, LH, TSH, free thyroxine, prolactin, testosterone; semen characteristics	No statistically significant differences in FSH, LH, prolactin, testosterone, thyroid measures between low- and high-urinary-uranium groups Semen characteristics: Total sperm count [≥40 million] Low urinary uranium, 286.6 ± 44.8 million; high urinary uranium, 583.5 ± 106.1 million; $p = 0.02$	Prescription psychotropic-, antidepressant-drug use

				Total progressive sperm (WHO Class A and B) [≥20 million] Low urinary uranium, 108.2 ± 19.2 million; high urinary uranium, 220.9 ± 44.0 million; $p = 0.03$ Total rapid progressive sperm (WHO Class A) [≥10 million] Low urinary uranium, 81.3 ± 15.4 million; high urinary uranium, 155.5 ± 31.1 million; $p = 0.04$
McDiarmid et al., 2004 Case series	39 Gulf War veterans exposed to DU during friendly-fire incidents in February 1991, examined in April-July 2001, followup 1994-2001	Same exposure as in McDiarmid et al., 2001	Neuroendocrine measures: FSH, LH, prolactin, TSH, free thyroxine, testosterone; semen characteristics	No statistically significant differences in reproductive-health measures
McDiarmid et al., 2006 Case series	32 Gulf War veterans exposed to DU during friendly-fire incidents, examined in April-July 2003	Same exposure as in McDiarmid et al., 2001	Neuroendocrine measures: FSH, LH, prolactin, TSH, free thyroxine, testosterone; semen characteristics	No statistically significant differences in reproductive-health measures

Continued

TABLE 8-18 Continued

Study	Population	Exposure	Outcomes or Outcome Measures	Results	Adjustments
McDiarmid et al., 2007 Case series	34 Gulf War veterans exposed to DU during friendly-fire incidents, examined in April-June 2005	Exposure to DU by friendly fire, may have inhaled, ingested airborne DU particles, experienced wound contamination by DU; assessed urinary uranium concentration; both current and cumulative exposure measures reported	Neuroendocrine measures, semen characteristics	No statistically significant differences in reproductive-health measures	
Sumanovic-Glamuzina et al., 2003 Pre-post comparison	All liveborn, stillborn neonates in Maternity Ward of Mostar University Hospital of western Herzegovina, part of Bosnia and Herzegovina immediately (1995) and 5 years after (2000) 1991-1995 military activities	Living in western Herzegovina after military activities	Major congenital malformations	1995 cohort: Major malformations in 40 of 1,853 neonates (2.16%; 95% CI, 1.49-2.82%) 2000 cohort: Major malformations in 33 of 1,463 neonates (2.26%; 95% CI, 1.50-3.01%)	

NOTE: BDI = Beck Depression Inventory, CI = confidence interval, DU = depleted uranium, FSH = follicle-stimulating hormone, LH = luteinizing hormone, TSH = thyroid-stimulating hormone, WHO = World Health Organization.

REFERENCES

Archer, V. E., J. K. Wagoner, and F. E. Lundin, Jr. 1973. Cancer mortality among uranium mill workers. *Journal of Occupational Medicine* 15(1):11-14.

ATSDR (Agency for Toxic Substances and Disease Registry). 1999. *Toxicological profile for uranium.* Atlanta, GA: U.S. Department of Health and Human Services, Public Health Service.

Auvinen, A., P. Kurttio, J. Pekkanen, E. Pukkala, T. Ilus, and L. Salonen. 2002. Uranium and other natural radionuclides in drinking water and risk of leukemia: A case-cohort study in Finland. *Cancer Causes Control* 13(9):825-829.

Auvinen, A., L. Salonen, J. Pekkanen, E. Pukkala, T. Ilus, and P. Kurttio. 2005. Radon and other natural radionuclides in drinking water and risk of stomach cancer: A case-cohort study in Finland. *International Journal of Cancer* 114(1):109-113.

Beral, V., P. Fraser, L. Carpenter, M. Booth, A. Brown, and G. Rose. 1988. Mortality of employees of the atomic weapons establishment, 1951-82. *British Medical Journal* 297(6651):757-770.

Boiano, J. M., C. E. Moss, and G. A. Burr. 1989. *Health hazard evaluation report HETA 83-144-2001, feed materials production center*. Fernald, OH: Westinghouse Materials Company of Ohio.

Boice, J. D., Jr., W. L. Bigbee, M. T. Mumma, and W. J. Blot. 2003a. Cancer incidence in municipalities near two former nuclear materials processing facilities in Pennsylvania. *Health Physics* 85(6):678-690.

———. 2003b. Cancer mortality in counties near two former nuclear materials processing facilities in Pennsylvania, 1950-1995. *Health Physics* 85(6):691-700.

Boice, J. D., Jr., M. Mumma, S. Schweitzer, and W. J. Blot. 2003c. Cancer mortality in a Texas county with prior uranium mining and milling activities, 1950-2001. *Journal of Radiological Protection* 23(3):247-262.

Boice, J. D., S. S. Cohen, M. T. Mumma, E. Dupree Ellis, K. F. Eckerman, R. W. Leggett, B. B. Boecker, A. B. Brill, and B. E. Henderson. 2006. Mortality among radiation workers at Rocketdyne (Atomics International), 1948-1999. *Radiation Research* 166(1 Pt 1):98-115.

Brown, D. P., and T. Bloom. 1987. *Mortality among uranium enrichment workers*. Cincinnati, OH: NIOSH.

Checkoway, H., N. Pearce, D. J. Crawford-Brown, and D. L. Cragle. 1988. Radiation doses and cause-specific mortality among workers at a nuclear materials fabrication plant. *American Journal of Epidemiology* 127(2):255-266.

Cragle, D. L., R. W. Mclain, J. R. Qualters, G. S. Wilikinson, W. G. Tankersley, and C. C. Lushbaugh. 1988. Mortality among workers at a nuclear fuels production facility. *American Journal of Industrial Medicine* 14:379-401.

Cross, F. T., R. F. Palmer, R. H. Busch, R. E. Filipy, and B. O. Stuart. 1981. Development of lesions in Syrian golden hamsters following exposure to radon daughters and uranium ore dust. *Health Physics* 41(1):135-153.

Dupree, E. A., D. L. Cragle, R. W. McLain, D. J. Crawford-Brown, and M. J. Teta. 1987. Mortality among workers at a uranium processing facility, the Linde air products company ceramics plant, 1943-1949. *Scandinavian Journal of Work, Environment, and Health* 13(2):100-107.

Dupree, E. A., J. P. Watkins, J. N. Ingle, P. W. Wallace, C. M. West, and W. G. Tankersley. 1995. Uranium dust exposure and lung cancer risk in four uranium processing operations. *Epidemiology* 6(4):370-375.

Dupree-Ellis, E., J. Watkins, J. N. Ingle, and J. Phillips. 2000. External radiation exposure and mortality in a cohort of uranium processing workers. *American Journal of Epidemiology* 152(1):91-95.

Filippova, L. G., A. P. Nifatov, and E. R. Liubchanskii. 1978. Some of the long-term sequelae of giving rats enriched uranium. *Radiobiologiia* 18(3):400-405.

Frome, E. L., D. L. Cragle, and R. W. McLain. 1990. Poisson regression analysis of the mortality among a cohort of World War II nuclear industry workers. *Radiation Research* 123(2):138-152.

Frome, E. L., D. L. Cragle, J. P. Watkins, S. Wing, C. M. Shy, W. G. Tankersley, and C. M. West. 1997. A mortality study of employees of the nuclear industry in Oak Ridge, Tennessee. *Radiation Research* 148(1):64-80.

Gustavsson, P., M. Talback, A. Lundin, B. Lagercrantz, P. E. Gyllestad, and L. Fornell. 2004. Incidence of cancer among Swedish military and civil personnel involved in UN missions in the Balkans 1989-99. *Occupational and Environmental Medicine* 61(2):171-173.

Hadjimichael, O. C., A. M. Ostfeld, D. A. D'Atri, and R. E. Brubaker. 1983. Mortality and cancer incidence experience of employees in a nuclear fuels fabrication plant. *Journal of Occupational Medicine* 25(1):48-61.

Hahn, F. F., R. A. Guilmette, and M. D. Hoover. 2002. Implanted depleted uranium fragments cause soft tissue sarcomas in the muscles of rats. *Environmental Health Perspectives* 110(1):51-59.

IOM (Institute of Medicine). 2000. *Gulf War and health, volume 1: Depleted uranium, pyridostigmine bromide, sarin, vaccines*. Washington, DC: National Academy Press.

———. 2006. *Gulf War and health, volume 4: Health effects of serving in the Gulf War*. Washington, DC: The National Academies Press.

Kathren, R. L., J. F. McInroy, R. H. Moore, and S. E. Dietert. 1989. Uranium in the tissues of an occupationally exposed individual. *Health Physics* 57(1):17-21.

Kurttio, P., A. Auvinen, L. Salonen, H. Saha, J. Pekkanen, I. Makelainen, S. B. Vaisanen, I. M. Penttila, and H. Komulainen. 2002. Renal effects of uranium in drinking water. *Environmental Health Perspectives* 110(4):337-342.

Kurttio, P., H. Komulainen, A. Leino, L. Salonen, A. Auvinen, and H. Saha. 2005. Bone as a possible target of chemical toxicity of natural uranium in drinking water. *Environmental Health Perspectives* 113(1):68-72.

Kurttio, P., A. Harmoinen, H. Saha, L. Salonen, Z. Karpas, H. Komulainen, and A. Auvinen. 2006a. Kidney toxicity of ingested uranium from drinking water. *American Journal of Kidney Disease* 47(6):972-982.

Kurttio, P., L. Salonen, T. Ilus, J. Pekkanen, E. Pukkala, and A. Auvinen. 2006b. Well water radioactivity and risk of cancers of the urinary organs. *Environmental Research* 102(3):333-338.

Leach, L. J., C. L. Yuile, H. C. Hodge, G. E. Sylvester, and H. B. Wilson. 1973. A five-year inhalation study with natural uranium dioxide (UO_2) dust. II. Postexposure retention and biologic effects in the monkey, dog and rat. *Health Physics* 25(3):239-258.

Loomis, D. P., and S. H. Wolf. 1996. Mortality of workers at a nuclear materials production plant at Oak Ridge, Tennessee, 1947-1990. *American Journal of Industrial Medicine* 29(2):131-141.

Macfarlane, G. J., A. M. Biggs, N. Maconochie, M. Hotopf, P. Doyle, and M. Lunt. 2003. Incidence of cancer among UK Gulf War veterans: Cohort study. *British Medical Journal* 327(7428):1373.

Macfarlane, G. J., M. Hotopf, N. Maconochie, N. Blatchley, A. Richards, and M. Lunt. 2005. Long-term mortality amongst Gulf War veterans: Is there a relationship with experiences during deployment and subsequent morbidity? *International Journal of Epidemiology* 34(6):1403-1408.

Maynard, E. A., and H. C. Hodge. 1949. Studies of the toxicity of various uranium compounds when fed to experimental animals. In *Pharmacology and toxicology of uranium compounds*, edited by C. Voegtlin and H. C. Hodge. Pp. 309-376.

McDiarmid, M. A., J. P. Keogh, F. J. Hooper, K. McPhaul, K. Squibb, R. Kane, R. DiPino, M. Kabat, B. Kaup, L. Anderson, D. Hoover, L. Brown, M. Hamilton, D. Jacobson-Kram, B. Burrows, and M. Walsh. 2000. Health effects of depleted uranium on exposed Gulf War veterans. *Environmental Research* 82(2):168-180.

McDiarmid, M. A., K. Squibb, S. Engelhardt, M. Oliver, P. Gucer, P. D. Wilson, R. Kane, M. Kabat, B. Kaup, L. Anderson, D. Hoover, L. Brown, and D. Jacobson-Kram. 2001. Surveillance of depleted uranium exposed Gulf War veterans: Health effects observed in an enlarged "friendly fire" cohort. *Journal of Occupational and Environmental Medicine* 43(12):991-1000.

McDiarmid, M. A., F. J. Hooper, K. Squibb, K. McPhaul, S. M. Engelhardt, R. Kane, R. DiPino, and M. Kabat. 2002. Health effects and biological monitoring results of Gulf War veterans exposed to depleted uranium. *Military Medicine* 167(2 Suppl):123-124.

McDiarmid, M. A., S. Engelhardt, M. Oliver, P. Gucer, P. D. Wilson, R. Kane, M. Kabat, B. Kaup, L. Anderson, D. Hoover, L. Brown, B. Handwerger, R. J. Albertini, D. Jacobson-Kram, C. D. Thorne, and K. S. Squibb. 2004. Health effects of depleted uranium on exposed Gulf War veterans: A 10-year follow-up. *Journal of Toxicology and Environmental Health A* 67(4):277-296.

McDiarmid, M. A., S. M. Engelhardt, M. Oliver, P. Gucer, P. D. Wilson, R. Kane, M. Kabat, B. Kaup, L. Anderson, D. Hoover, L. Brown, R. J. Albertini, R. Gudi, D. Jacobson-Kram, C. D. Thorne, and K. S. Squibb. 2006. Biological monitoring and surveillance results of Gulf War I veterans exposed to depleted uranium. *International Archives of Occupational and Environmental Health* 79(1):11-21.

McDiarmid, M. A., S. M. Engelhardt, M. Oliver, P. Gucer, P. D. Wilson, R. Kane, A. Cernich, B. Kaup, L. Anderson, D. Hoover, L. Brown, R. Albertini, R. Gudi, D. Jacobson-Kram, and K. S. Squibb. 2007. Health surveillance of Gulf War I veterans exposed to depleted uranium: Updating the cohort. *Health Physics* 93(1):60-73.

McGeoghegan, D., and K. Binks. 2000a. The mortality and cancer morbidity experience of workers at the Capenhurst uranium enrichment facility 1946-95. *Journal of Radiological Protection* 20(4):381-401.

———. 2000b. The mortality and cancer morbidity experience of workers at the Springfields uranium production facility, 1946-95. *Journal of Radiological Protection* 20(2):111-137.

———. 2001. The mortality and cancer morbidity experience of employees at the Chapelcross plant of British Nuclear Fuels PLC, 1955-95. *Journal of Radiological Protection* 21(3):221-250.

Miller, A. C., C. Bonait-Pellie, R. F. Merlot, J. Michel, M. Stewart, and P. D. Lison. 2005. Leukemic transformation of hematopoietic cells in mice internally exposed to depleted uranium. *Molecular and Cellular Biochemistry* 279(1-2):97-104.

Mitchel, R. E. J., J. S. Jackson, and B. Heinmiller. 1999. Inhaled uranium ore dust and lung cancer risk in rats. *Health Physics* 76(2):145-155.

NRC (National Research Council). 2008. *Review of toxicologic and radiologic risks to military personnel from exposure to depleted uranium during and after combat*. Washington, DC: The National Academies Press.

Nuccetelli, C., M. Grandolfo, and S. Risica. 2005. Depleted uranium: Possible health effects and experimental issues. *Microchemical Journal* 79(1-2):331-335.

Pinkerton, L. E., T. F. Bloom, M. J. Hein, and E. M. Ward. 2004. Mortality among a cohort of uranium mill workers: An update. *Occupational and Environmental Medicine* 61(1):57-64.

Pinney, S. M., R. W. Freyberg, G. E. Levine, D. E. Brannen, L. S. Mark, J. M. Nasuta, C. D. Tebbe, J. M. Buckholz, and R. Wones. 2003. Health effects in community residents near a uranium plant at Fernald, Ohio, USA. *International Journal of Occupational Medicine and Environmental Health* 16(2):139-153.

Polednak, A. P., and E. L. Frome. 1981. Mortality among men employed between 1943 and 1947 at a uranium-processing plant. *Journal of Occupational Medicine* 23(3):169-178.

Richardson, D. B., and S. Wing. 2006. Lung cancer mortality among workers at a nuclear materials fabrication plant. *American Journal of Industrial Medicine* 49(2):102-111.

Ritz, B. 1999. Radiation exposure and cancer mortality in uranium processing workers. *Epidemiology* 10(5):531-538.

Ritz, B., H. Morgenstern, D. Crawford-Brown, and B. Young. 2000. The effects of internal radiation exposure on cancer mortality in nuclear workers at Rocketdyne/Atomics International. *Environmental Health Perspectives* 108(8):743-751.

Shawky, S., H. A. Amer, M. I. Hussein, Z. el-Mahdy, and M. Mustafa. 2002. Uranium bioassay and radioactive dust measurements at some uranium processing sites in Egypt—health effects. *Journal of Environmental Monitoring* 4(4):588-591.

Stayner, L. T., T. Meinhardt, R. Lemen, D. Bayliss, R. Herrick, G. R. Reeve, A. B. Smith, and W. Halperin. 1985. A retrospective cohort mortality study of a phosphate fertilizer production facility. *Archives of Environmental Health* 40(13):133-138.

Storm, H. H., H. O. Jorgensen, A. M. Kejs, and G. Engholm. 2006. Depleted uranium and cancer in Danish Balkan veterans deployed 1992-2001. *European Journal of Cancer* 42(14):2355-2358.

Sumanovic-Glamuzina, D., V. Saraga-Karacic, Z. Roncevic, A. Milanov, T. Bozic, and M. Boranic. 2003. Incidence of major congenital malformations in a region of Bosnia and Herzegovina allegedly polluted with depleted uranium. *Croatian Medical Journal* 44(5):579-584.

USACHPPM (U.S. Army Center for Health Promotion and Preventive Medicine). 2004. *Capstone report: Depleted uranium aerosol doses and risks: Summary of U.S. Assessments*. Fort Belvoir, VA: U.S. Army Heavy Metals Office, Chemical and Biological Defense Information Analysis Center.

Wagoner, J. K., V. E. Archer, B. E. Carroll, D. A. Holaday, and P. A. Lawrence. 1964. Cancer mortality patterns among U.S. uranium miners and millers, 1950 through 1962. *Journal of the National Cancer Institute* 32(4):787-801.

Waxweiler, R. J., V. E. Archer, R. J. Roscoe, A. Watanabe, and M. J. Thun. 1983. Mortality patterns among a retrospective cohort of uranium mill workers. In Epidemiology Applied to Health Physics, Proceedings of the Sixteenth Midyear Topical Meeting of the Health Physics Society, Albuquerque, New Mexico, January 9-13:428-435.

Index

A

B

T

U

V

W